# CALCULATED RISKS

## The Toxicity and Human Health Risks of Chemicals in Our Environment

### Second Edition, with major revisions and additions

Safeguarding economic prosperity, whilst protecting human health and the environment, is at the forefront of scientific and public interest. This book provides a practical and balanced view on toxicology, control, risk assessment, and risk management, addressing the interplay between science and public health policy. This fully revised and updated new edition provides a detailed analysis on chemical and by-product exposure, how they enter the body and the suitability of imposed safety limits. New chapters on dose, with particular emphasis on children and vulnerable subpopulations, reproductive and developmental toxicants, and toxicity testing are included. With updated and comprehensive coverage of international developments in risk management and safety, this will have broad appeal to researchers and professionals involved in chemical safety and regulation, as well as to the general reader interested in environmental pollution and public health.

JOSEPH V. RODRICKS was a founding principal of ENVIRON International Corporation, a consultancy firm on environmental and health issues. Since 1980 he has consulted for many corporations and institutions, including the World Health Organization, and in 2005 he received the Outstanding Practitioner Award from the Society for Risk Analysis. The first edition of *Calculated Risks* won an "Honourable Mentions" award from the American Medical Writers Association.

# Praise for first edition

"Calculated Risks demystifies the science and policies of risk assessment. It has become a staple in risk education, and is essential reading for students and professionals in public health, environmental protection, and public policy."     Thomas A. Burke, Professor and Associate Chair, Bloomberg School of Public Health, Johns Hopkins University

". . . Rodricks has made the difficult topic of risk assessment accessible to the regulatory, policy and scientific communities. Calculated Risks focuses on the science of assessing health risks and provides a framework for understanding this complex topic. It should be required reading for those concerned about environmental pollution and protection of health and environment."     Carol J. Henry, Vice President Science and Research, American Chemistry Council

". . . presents a practical and balanced clarification of the scientific basis for our concerns and uncertainties. It should serve to refocus the debate." *Biology Digest*

". . . provides access to the science and uncertainty behind the oft-quoted risks of toxic chemicals . . . The reader who completes the book is likely to know much more about the limitations of all assessments of risk." *BioScience*

"Rather than attempting to expose governmental and corporate ignorance, negligence or corruption, this book explores the underlying scientific issues. It presents a clarification of the scientific basis for our concerns and uncertainties." *The Bulletin of Science, Technology and Society*

". . . a well-organized and readable text . . . The book should be recommended reading for those interested in obtaining an understanding of risk assessment." *Canadian Field Naturalist*

"It is difficult to praise this book enough. An evenhanded text that emphasizes complexity and reveals the gaps in our knowledge rather than oversimplifying the science of toxicology, *Calculated Risks: The Toxicity and Human Health Hazards of Chemicals in Our Environment* belongs on the shelves of every environmental organization. Writing in a manner that neither condescends nor baffles his readers, Joseph V. Rodricks has produced a text that if used as a point of departure in discussing siting, pollution, and similar disputes could save time and effort . . . This book is the basic text we all should read." *Environment*

". . . the best book we have yet seen on the theory of risk assessment – lucidly written, and evenhanded . . . If you want to understand the theory of risk assessment from the viewpoint of a successful risk assessor, this is the book for you." *Rachel's Hazardous Waste News*

# CALCULATED RISKS

## The Toxicity and Human Health Risks of Chemicals in Our Environment

Second Edition, with major revisions and additions

Joseph V. Rodricks

CAMBRIDGE
UNIVERSITY PRESS

CAMBRIDGE UNIVERSITY PRESS
Cambridge, New York, Melbourne, Madrid, Cape Town, Singapore, São Paulo

Cambridge University Press
The Edinburgh Building, Cambridge CB2 2RU, UK

Published in the United States of America by Cambridge University Press, New York

www.cambridge.org
Information on this title: www.cambridge.org/9780521783088

© 1992, 2007

First edition published 1992
Second edition published 2007

Printed in the United Kingdom at the University Press, Cambridge

*A catalog record for this publication is available from the British Library*

ISBN-13  978-0-521-78308-8 hardback
ISBN-10  0-521-78308-9 hardback

# Contents

# Preface to the first edition

Think how many carcinogens are household names: asbestos, cigarette smoke (a mixture of several thousand chemical compounds), DES, dioxin, saccharin, arsenic, PCBs, radon, EDB, Alar. Hundreds more of these substances, some very obscure, are known to the scientific and medical community, and many of these are scattered throughout the land at thousands of hazardous waste sites similar to Love Canal. People are exposed to these dreadful substances through the air they breathe, the water they drink and bathe in, and the foods they eat. Chemicals can also produce many other types of health damage, some very serious, such as birth defects and damage to our nervous and immune systems.

The chemical accident at Bhopal, India, in late 1984, is only the worst example of events that take place almost daily, on a smaller scale, throughout the world. Human beings are not the only potential victims of chemical toxicity – all of life on earth can be affected. Chemicals are ravaging human health and the environment, and conditions are worsening.

But wait. Let's remember that chemicals have virtually transformed the modern world in extraordinarily beneficial ways. During the past 100 years the chemical industry has offered up, and we have eagerly consumed, thousands of highly useful materials and products. Among these products are many that have had profoundly beneficial effects on human health – antibiotics and other remarkable medicinal agents to prevent and cure diseases, pesticides to protect crops, preservatives to protect the food supply, plastics, fibers, metals and hundreds of other materials that have enhanced the safety and pleasures of modern life. Perhaps the misuse of certain chemicals has caused some small degree of harm, but on balance the huge benefits of modern chemical society clearly outweigh the exceedingly small risks these products may carry. Moreover, we have made and are continuing to make progress in controlling the risks of chemical technology.

Somewhere between these two views sits a somewhat befuddled scientific and medical community, attempting to sort the true from the false, and not quite sure how it should respond to the public while this sorting takes place. What science can now say with reasonably high certainty about the risks of chemical technology falls far short of the knowledge about those risks that our citizens are seeking. And, although substantial progress in scientific understanding has been made during the past three to four decades, it will probably be another several decades before the questions about chemical risks facing us today can be answered with the degree of certainty normally sought by scientists.

It is not at all surprising that confusion and controversy should arise when knowledge is absent or weak. When, as in the case of the risks of the products and byproducts of chemical technology, scientists know just enough to raise fearful suspicions, but do not always know enough to separate the true fears from the false, other social forces take command. Among the most important of these forces are the environmental laws that sometimes require regulatory authorities to act even before scientific understanding is firm. When the consequences of these actions cause economic harm, combat begins. Depending on which side of the battle one sits, fears about chemical risks are emphasized or downplayed. The form of the battle that will occur following what have become routine announcements about carcinogens in pancakes or apples, or nervous system poisons in drinking water or soft drinks, is now highly predictable. Except for a few brave (or foolish?) souls, the scientific community tends to remain relatively impassive in such circumstances, at most calling for "more research." Those scientists who are sufficiently intrepid to offer opinions tend to be scorned either as environmentalist quacks or industry hacks, who have departed from the traditional, scientifically acceptable standards of proof. Perhaps they have, but as we shall see, there is certainly an argument to be made on their behalf.

The question of whether and to what degree chemicals present in air, food, drinking water, medicinal agents, consumer products, and in the work place pose a threat to human health is obviously of enormous social and medical importance. This book is an attempt to answer this question with as much certainty as science can currently offer. It is in part a book of popular science – that is, it attempts to provide for the layman a view of the sciences of toxicology and chemical carcinogenesis (considered by some a branch of toxicology). It describes how the toxic properties of chemicals are identified, and how scientists make judgments about chemical risks. What is known with reasonable certainty is separated from the speculative; the large gray areas of science falling between these extremes are also sketched out. Toxicology, the science of poisons, is such a rich and fascinating subject, that it deserves more widespread recognition on purely intellectual grounds. Because it is now such an important tool in public health and regulatory decision-making, it is essential that its elements be widely understood.

The focus of this book is on the methods and principles of toxicology and risk assessment, and not on particular toxic agents or on the scientists who have built the discipline. To emphasize specific agents and scientists would have resulted in too great a departure from the book's second aim – to cast a little light upon the difficult interaction between science and the development of public health and regulatory policies. What is of interest here is not the administrative detail of policy implementation, which can be a rather unlively topic, but the principles that have come to govern the interaction of a highly uncertain scientific enterprise with the social demand for definitive actions regarding matters of public health.

The purpose of this book is, then, to describe and to clarify the scientific reasons for our present concerns about chemicals in the environment; the strengths and weaknesses of our scientific understanding; and the interplay between science and public policy. Unlike most other works related to these subjects, it is not an attempt to expose governmental and corporate ignorance, negligence or corruption. There is no end to literature on this subject, much of it presenting an incomplete or biased view of current scientific understanding of the effects of chemicals on human health and the environment. Perhaps a little clarification of the scientific bases for our concerns and the uncertainties that accompany them, and of the dilemmas facing decision-makers, will serve to refocus and advance the debate.

A word about organization of topics is in order. First, it is important to understand what we mean when we talk about "chemicals." Many people think the term refers only to generally noxious materials that are manufactured in industrial swamps, frequently for no good purpose. The existence of such an image impedes understanding of toxicology and needs to be corrected. Moreover, because the molecular architecture of chemicals is a determinant of their behavior in biological systems, it is important to create a little understanding of the principles of chemical structure and behavior. For these reasons, we begin with a brief review of some fundamentals of chemistry.

The two ultimate sources of chemicals – nature and industrial and laboratory synthesis – are then briefly described. This review sets the stage for a discussion of how human beings become exposed to chemicals. The conditions of human exposure are a critical determinant of whether and how a chemical will produce injury or disease, so the discussion of chemical sources and exposures naturally leads to the major subject of the book – the science of toxicology.

The major subjects of the last third of this volume are risk assessment – the process of determining the likelihood that chemical exposures have or will produce toxicity – and risk control, or management, and the associated topic of public perceptions of risk in relation to the judgments of experts. It is particularly in these areas that the scientific uncertainties become most visible and the public debate begins to heat up. The final chapter contains some suggestions for improving the current state-of-affairs and also sets out some new challenges. Risks to human health are the subject of this book; risks to the rest of the living and non-living environment are not covered. The absence of this topic from the present volume has only to do with the author's interests and knowledge and says nothing about its relative importance.

Much of the discussion of risk assessment turns on the activities of regulatory agencies responsible for enforcing the two dozen or so federal laws calling for restrictions of one sort or another on human exposures to environmental chemicals. The Environmental Protection Agency (EPA) has responsibilities for air and water pollutants, pesticides, hazardous wastes, and industrial chemicals not covered by other statutes. The Food and Drug Administration (FDA), part of the Department of Health and Human Services, manages risks from foods and substances added thereto, drugs for both human and veterinary uses (some of the latter can reach people through animal products such as meat, milk and eggs), cosmetics, and constituents of medical devices. The Occupational Safety and Health Administration (OSHA), a unit of the Labor Department, handles chemical exposures in the workplace. Consumer products not covered by other agencies fall to the Consumer Product Safety Commission (CPSC). Other agencies with similar, though somewhat narrower responsibilities, include the

Food Safety and Inspection Service of the Department of Agriculture (for meat, poultry, and eggs) and the Department of Transportation. Although laws and regulatory programs vary, most countries have agencies with similar sets of responsibilities.

The use of the phrase "regulatory risk assessments" throughout this book may seem odd, because risk assessment is a scientific activity and its conduct, it would seem, should be independent of where it is undertaken. But we shall see that a scientific consensus on the proper conduct of risk assessment does not exist, and regulatory agencies have had to adopt, as a matter of policy, certain assumptions that do not have universal acceptance in the scientific community. The agencies do this to allow them to operate in accordance with their legal mandates, and one of the purposes of this book is to create understanding of (but not necessarily to urge agreement with) these regulatory policies.

Before embarking on what is perhaps an overly systematic approach to our subject, we should attempt to develop a bird's-eye view of the entire landscape. We shall use a specific example – the case of a group of chemicals called aflatoxins – to illustrate the type of problem this book is designed to explore.

# Preface to the second edition

The central topics of this book have, since its publication in 1992, become permanent occupants of the public health agendas of governments everywhere. The imposition of controls on the production, uses, and environmental releases of chemical products, by-products, and environmental pollutants, whatever their sources, has become the ambition of most societies, and many now claim to rely upon risk assessments to guide their decisions about the necessity for, and extent of, such controls. Indeed, the risk assessment framework sketched out in the first edition, and given far more extensive treatment in this one, has come to be seen as a most powerful tool for evaluating and putting into useful form the complex, diverse, often inconsistent, and always incomplete scientific information and knowledge we have been able to accumulate about the health hazards and risks all chemicals pose if exposures become excessive. So in this edition the reader will be provided with a broader and deeper look at risk assessment as it continues to evolve as a scientific enterprise, and in its role as the bridge between basic and applied research and the many forms of decision-making aimed at risk reduction. Chapters dealing with the early evolution of the risk assessment framework and the several principles and concepts that gave rise to its structure, the ways in which relevant scientific data and knowledge are put to use within that framework, and some of the new challenges risk assessment faces, are almost wholly new in this edition. This much expanded look at the many dimensions and evolving uses of the risk assessment framework was driven in part by the comments I received from students and many other readers that the first edition had been too limited on these topics. So I hope I have now delivered something more satisfying.

The new edition is otherwise arranged much like the first, beginning as it does with a discussion of the world of chemicals and moving systematically to the subjects of human exposures to these substances and the various ways in which they can cause harm if those exposures somehow become excessive. Just how we come to know how much is excessive for an individual substance is given much discussion. The sciences of toxicology and epidemiology, their methods and applications, are extensively treated, in updated and almost wholly rewritten chapters. All the major types of toxic harm, including cancer, are reviewed. New issues – everything from chemicals commonly released during industrial accidents, some used for terrorist purposes, and some of the recently uncovered sources of chronic toxicity can be found throughout the new edition. New thinking on mechanisms of toxic action and the use of mechanistic information in risk assessment comes up time and again. Persistent chemicals, chemicals found in the human body, endocrine-disrupting substances, and products of the hottest new technology – that carried out at the nano scale – are here. Expanding uses of the risk assessment framework include the problems of nutrient deficiency and excess. Microbial pathogens that cause food poisoning are given some space in the risk assessment chapter dealing with new challenges.

I have attempted to hold to a writing style that is accessible to the non-professional and that is at the same time at least moderately interesting to professionals. The problem of keeping professionals interested is eased a little by the fact that so many different areas of the health sciences, both basic and applied, medicine, and the environmental sciences are drawn into the subjects I cover; and I hope a reading of this book provides a satisfyingly broad perspective.

At the beginning of this preface I described an expanding web of risk assessment practitioners and users; this description, while accurate, may create a false impression. It is true that the subjects of this book and the ways in which they are brought together within the risk assessment framework are increasingly discussed and advocated, by governments and by many non-governmental organizations and corporations. Almost everywhere risk assessment is promoted as the guide to a safer environment. The subject is increasingly taught in formal and informal settings all over the world. This is admirable.

What is not so admirable is that in so many of these institutions there is a lack of commitment to turning discussion and advocacy into action. Lack of resources to develop and implement risk-based health protection programs explains much but not all of the inaction. In fact, once it is recognized that the practice of risk assessment does not generate new information and knowledge – that can only come from research – but that it serves rather to organize and tell us what we know and do not know about threats to our health, then it becomes clear that support for risk assessment-based decision-making requires that we support the scientific research, epidemiology investigations, and toxicity testing upon which it depends. Supporting risk assessment without at the same time providing the much more substantial resources needed to support research and testing becomes something of an empty gesture.

Even when resources are available to produce reliable assessments, inaction may result when the results of these assessments put risk managers into politically awkward positions. And, of course, many risk assessments are produced by well-meaning but inexperienced, inadequately trained individuals, and one would hope risk managers would have mechanisms in place to eliminate incompetent work.

Even admitting these various impediments to the broader use of risk-based decisions to protect the public health, much progress has been made, and there is no reason to believe this trend will not continue. I hope this new edition contributes in some way to this trend.

I am grateful beyond words to the technical guidance provided by my long-term associate, Duncan Turnbull, and to all manner of assistance provided by Gail Livingston. My dear wife Karen Hulebak, herself a public health scientist, kept me from veering off-track at many points; any such veering that remains in the book is due to my own lack of control. The Second Edition is dedicated to my daughter, Elizabeth.

# Abbreviations

| | |
|---|---|
| ACGIH | American Conference of Governmental Industrial Hygienists |
| ATSDR | Agency for Toxic Substances and Disease Registry (DHHS) |
| CDC | Centers for Disease Control (DHHS) |
| CPSC | Consumer Product Safety Commission |
| EPA | Environmental Protection Agency |
| FAO | Food and Agriculture Organization (UN) |
| FDA | Food and Drug Administration (DHHS) |
| FSIS | Food Safety and Inspection Service (USDA) |
| IARC | International Agency for Research on Cancer (WHO) |
| NCI | National Cancer Institute (DHHS) |
| NIEHS | National Institutes of Environmental Health Sciences |
| NIH | National Institutes of Health (DHHS) |
| NIOSH | National Institute of Occupational Safety and Health (DHHS) |
| NTP | National Toxicology Program (DHHS) |
| OSHA | Occupational Safety and Health Administration (DOL) |
| WHO | World Health Organization (UN) |
| DHHS | Department of Health and Human Services |
| DOL | Department of Labor |
| UN | United Nations |
| USDA | United States Department of Agriculture |
| | |
| ADI | allowable daily intake |
| ADME | absorption, distribution, metabolism, elimination |

| | |
|---|---|
| AEGL | acute exposure guideline |
| BMD | benchmark dose |
| CI | confidence interval |
| LOAEL | lowest observed adverse effect level |
| MCL | maximum contaminant level |
| MCLG | maximum contaminant level goal |
| MRL | minimum risk level |
| NOAEL | no observed adverse effect level |
| OR | odds-ratio |
| PD | pharmacodynamics |
| PEL | permissible exposure level |
| PK | pharmacokinetics |
| POD | point-of-departure |
| RfC | toxicity reference concentration |
| RfD | toxicity reference dose |
| RR | relative risk |
| TDI | tolerable daily intake |
| TLV | threshold limit value[1] |
| UF | uncertainty factor |
| UL | upper level |

[1] Registered Trademark of the American Conference of Governmental Industrial Hygienists (ACGIH).

# Prologue – groundnuts, cancer, and a small red book

In the fall of 1960 thousands of turkey poults and other animals started dying throughout southern England. Veterinarians were at first stymied about the cause of what they came to label "Turkey X disease," but because so many birds were affected, a major investigation into its origins was undertaken. In 1961 a report from three scientists at London's Tropical Products Institute and a veterinarian at the Ministry of Agriculture's laboratory at Weybridge, entitled "Toxicity associated with certain samples of groundnuts," was published in the internationally prominent scientific journal, *Nature*. Groundnuts, as everyone in America knows, are actually peanuts, and peanut meal is an important component of animal feed. It appeared that the turkeys had been poisoned by some agent present in the peanut meal component of their feed. The British investigators found that the poisonous agent was not a component of the peanuts themselves, but was found only in peanuts that had become contaminated with a certain mold.

It also became clear that the mold itself – identified by the mold experts (mycologists) as the fairly common species *Aspergillus flavus* – was not directly responsible for the poisoning. Turkey X disease could be reproduced in the laboratory not only when birds were fed peanut meal contaminated with living mold, but also when fed the same meal after the mold had been killed.

Chemists have known for a long time that molds are immensely productive manufacturers of organic chemical agents. Perhaps the best known mold product is penicillin, but this is only one of thousands of such products that can be produced by molds. Why molds are so good at chemical synthesis is not entirely clear, but they surely can produce an array of molecules whose complexities are greatly admired by the organic chemist.

In fact Turkey X disease was by no means the first example of a mold-related poisoning. Both the veterinary and public health literature contain hundreds of references to animal and human poisonings associated with the consumption of feeds or foods that had molded, not only with *Aspergillus flavus*, but also with many other mold species. Perhaps the largest outbreaks of human poisonings produced by mold toxins occurred in areas of the Soviet Union just before and during the Second World War. Cereal grains left in the fields over the winter, for lack of sufficient labor to bring them in, became molded with certain varieties that grow especially well, and produce their toxic products, in the cold and under the snow. Consumption of molded cereals in the following springtime led to massive outbreaks of human poisonings characterized by hemorrhaging and other dreadful effects. The Soviet investigators dubbed the disease alimentary toxic aleukia (ATA). The mold chemicals, or *mycotoxins* ("myco" is from the Greek word for fungus, mykes), responsible for ATA are now known to fall into a class of extremely complex organic molecules called trichothecenes, although toxicologists are still at work trying to reconstruct the exact causes of this condition. Veterinary, but probably not human, poisonings with this class of mycotoxins still take place in several areas of the world.

Even older than ATA is ergotism. Ergot poisoning was widespread in Europe throughout the Middle Ages, and has occurred episodically on a smaller scale many times since. The most notable recent outbreak occurred in France in 1951. This gruesome intoxication is produced by chemical products of *Claviceps purpurea*, a purple-colored mold that grows especially well on rye, wheat, and other grains. Most of the ergot chemicals are in a class called alkaloids, one member of which can be easily modified to produce the hallucinogenic agent, LSD, which of course came into popular use as a recreational drug during the 1960s. Ergot poisons produce a wide spectrum of horrible effects, including extremely painful convulsions, blindness, and gangrene. Parts of the body afflicted with gangrenous lesions blacken, shrink, dry up and may even fall off. The responsible mold is, unlike many others, fairly easy to spot, and normal care in the processing of grain into flour can eliminate the problem.

These and dozens more cases of mycotoxin poisonings were known to the investigators at the time they began delving into the causes of

Turkey X disease, so finding that a mold toxin was involved was no great surprise. But some new surprises were in store.

Investigations into the identity of the chemical agent responsible for Turkey X disease continued throughout the early 1960s at laboratories in several countries. At the Massachusetts Institute of Technology (MIT) a collaborative effort involving a group of toxicologists working under the direction of Gerald Wogan and a team of organic chemists headed by George Büchi had solved the mystery by 1965. The work of these scientists was a small masterpiece of the art of chemical and toxicological experimentation. After applying a long series of painstakingly careful extraction procedures to peanuts upon which the *Aspergillus flavus* mold had been allowed to grow, the research team isolated very small amounts of the substances that were responsible for the groundnut meal's poisonous properties. As is the custom among chemists, these substances were given a simple name that gave a clue to their source. Thus, from *Aspergillus flavus* toxin came the name aflatoxin.

Organic chemists are never satisfied with simply isolating and purifying such natural substances; their work is not complete until they identify the molecular structures of the substances they isolate. The case of aflatoxin presented a formidable challenge to the MIT team, because they were able to isolate only about 70 milligrams (mg) of purified aflatoxin with which to work (a milligram is one-thousandth of a gram, and a gram is about 1/30th of an ounce). But the team overcame this problem through a masterful series of experimental studies, and in 1965 published details about the molecular structure of aflatoxin. It is shown in Chapter 6 ("Identifying carcinogens").

It turned out that aflatoxin was actually a mixture of four different but closely related chemicals. All possessed the same molecular backbone of carbon, hydrogen, and oxygen atoms (which backbone was quite complex and not known to be present in any other natural or synthetic chemicals), but differed from one another in some minor details. Two of the aflatoxins emitted a blue fluorescence when they were irradiated with ultraviolet light, and so were named aflatoxin $B_1$ and $B_2$; the names aflatoxin $G_1$ and $G_2$ were assigned to the green-fluorescing compounds. The intense fluorescent properties of the aflatoxins would later prove an invaluable aid to chemists interested in measuring the amount of these substances present in various foods, because the intensity of the fluorescence was related to the amount of chemical present.

While all this elegant investigation was underway, it became clear that the aflatoxins were not uncommon contaminants of certain foods. A combination of the efforts of veterinarians investigating outbreaks of farm animal poisonings, survey work carried out by the Ministry of Agriculture in England, the US Department of Agriculture (USDA) and the Food and Drug Administration (FDA), and the investigations of individual scientists in laboratories throughout the world, revealed during the 1960s and 1970s that aflatoxins can be found fairly regularly in peanuts and certain peanut products, corn grown in certain geographical areas, and even in some varieties of tree nuts. Cottonseed grown in regions of the southwestern United States, but not in the southeast, was discovered to be susceptible. While peanut, corn, and cottonseed oils processed from contaminated products did not seem to carry the aflatoxins, these compounds did remain behind in the so-called "meals" made from these products. These meals are fed to poultry and livestock and, if they contain sufficiently high levels of aflatoxins, the chemical agents can be found in the derived food products – meat, eggs, and especially milk. The frequency of occurrence of the aflatoxins and the amounts found vary greatly from one geographical area to another, and seem to depend upon climate and agricultural and food storage practices.

While this work was underway, toxicologists were busy in several laboratories in the United States and Europe attempting to acquire a complete profile of aflatoxins' poisonous properties. These substances did seem to be responsible for several outbreaks of liver poisoning, sometimes resulting in death, in farm animals, but there was no evidence that aflatoxins reaching humans through various food products were causing similar harm. The most likely reason for this lack of evidence was the fact that the amounts of aflatoxins reaching humans through foods simply did not match the relatively large amounts that may contaminate animal feeds. Of course, if aflatoxins were indeed causing liver disease in people, it would be extremely difficult to find this out unless, as in the case of ATA or ergotism, the signs and symptoms were highly unusual and occurring relatively soon after exposure.

In experimental studies in laboratory settings, aflatoxins proved not only to be potent liver poisons, but also – and this was the great surprise – capable of producing malignant tumors, sometimes in great abundance, in rats, ferrets, guinea pigs, mice, monkeys, sheep, ducks, and rainbow trout (trout are exquisitely sensitive to aflatoxin-induced carcinogenicity). Several early studies from areas of the world in which human liver cancer rates are unusually high turned up evidence suggesting, but not clearly establishing, a role for the aflatoxins. Aflatoxin's cancer-producing properties were uncovered and reported in the scientific literature during the period 1961–1976, the same period during which these substances were discovered to be low-level but not infrequent contaminants of certain human foods.

What was to be done? Were the aflatoxins a real threat to the public health? How many cases of cancer could be attributed to them? Why was there no clear evidence that aflatoxins could produce cancers in exposed humans? How should we take into account the fact that the amounts of aflatoxins people might ingest through contaminated foods were typically very much less than the amounts that could be demonstrated experimentally to poison the livers of rodents, and to increase the rate of occurrence of malignancies in these several species? And if aflatoxins were indeed a public health menace, what steps should be taken to control or eliminate human exposure to them? Indeed, because aflatoxins occurred naturally, was it possible to control them at all?

These and other questions were much in the air during the decade from 1965 to 1975 at the Food and Drug Administration (FDA) – the public health and regulatory agency responsible for enforcing federal food laws and ensuring the safety of the food supply. Scientists and policy-makers from the FDA consulted aflatoxin experts in the scientific community, food technologists in affected industries, particularly those producing peanut, corn, and dairy products, and experts in agricultural practices. The agency decided that limits needed to be placed on the aflatoxin content of foods. In the 1960s, the FDA declared that peanut products containing aflatoxins in excess of 30 parts aflatoxin per billion parts of food (ppb) would be considered unfit for human consumption; a few years later the agency lowered the acceptable limit to 20 ppb. This ppb unit refers to the weight of aflatoxin divided by the weight of food; for one kilogram of peanut butter (about 2.2 lb), the 20 ppb limit restricts the aflatoxin content to 20 micrograms (one microgram is one-millionth of one gram – more will be said about these units later).

The FDA's decision was based on the conclusion that no completely safe level of human intake could be established for a cancer-causing chemical. This position led, in turn, to the position that if analytical chemists could be sure aflatoxins were present in a food, then the food could not be consumed without threatening human health. The question then was what is the smallest amount of aflatoxin that analytical chemists can reliably detect?: by 1968 this amount – or, more accurately, concentration – was 30 ppb, and because of improvements in analytical technology, the detection limit later dropped to 20 ppb. The analytical chemist dictated the FDA's position on acceptable aflatoxin limits.

It turned out that meeting a 20 ppb limit was not excessively burdensome on major manufacturers of peanut butter and other peanut products, at least in the United States; aflatoxin tended to concentrate in discolored or otherwise irregular peanuts, which, fortunately, could be picked up and rejected by modern electronic sorting machines. Manufacturers did, however, have to institute substantial additional quality control procedures to meet FDA limits, and many smaller manufacturers had trouble meeting a 20 ppb limit. An extensive USDA program of sampling and analysis of raw peanuts, which continues to this day, was also put into place as the first line of attack on the problem.

Did this FDA position make any scientific sense? It implied that if aflatoxin could be detected by reliable analysis, it was too risky to be consumed by humans, but that if the aflatoxin happened to be present below the minimum detectable concentration it was acceptable. (Analytical chemists can never declare that a chemical is *not* present. The best that can be done is to show that it is not present above some level – 20 ppb in the case of aflatoxins, and other, widely varying, levels in the case of other chemicals in the environment.) To be fair to the FDA, perhaps the word "acceptable" should be withdrawn; the agency's position was not so much that all concentrations of aflatoxin up to 20 ppb were acceptable, but that nothing much could be done about them, because the chemists could not determine whether they were truly present in a given lot of food until the concentration exceeded 20 ppb.

Was the FDA's position scientifically defensible? Let us offer two responses that might reflect the range of possible scientific opinion:

(1) Yes. FDA clearly did the right thing, and perhaps did not go far enough. Aflatoxins are surely potent cancer-causing agents in animals. We don't have significant human data, but this is very hard to get and we shouldn't wait for it before we institute controls. We know from much study that animal testing gives a reliable indication of human risk. We also know that cancer-causing chemicals are a special breed of toxicants – they can threaten health at any level of intake. We should therefore eliminate human exposure to such agents whenever we can, and, at the least, reduce exposure to the lowest possible level whenever we're not sure how to eliminate it.

(2) No. The FDA went too far. Aflatoxins can indeed cause liver toxicity in animals and are also carcinogenic. But they produce these adverse effects only at levels far above the FDA set limit. We should ensure some safety margin to protect humans, but 20 ppb is unnecessarily low and the policy that there is no safe level is not supported by scientific studies. Indeed, it's not even certain that aflatoxins represent a cancer risk to humans because animal testing is not known to be a reliable predictor of human risk. Moreover, the carcinogenic potency of aflatoxins varies greatly even among the several animal species in which they have been tested. Human evidence that aflatoxins cause cancer is unsubstantiated. There is no sound scientific basis for the FDA's position.

The whole matter of protective limits for aflatoxin became more complex in the early-to-mid 1970s when it became clear that analytical chemists could do far better than a 20 ppb detection limit. In several laboratories, aflatoxins could easily be detected as low as 5 ppb, and in some laboratories 1 ppb became almost routine. If the FDA was to follow a consistent policy, the agency would have had to call for these lower limits. But it did no such thing. It had become obvious to the FDA by the mid 1970s that a large fraction of the peanut butter produced by even the most technically advanced manufacturers would fail to meet a 1 ppb limit, and it was also apparent that other foods – corn meal and certain other corn products, some varieties of nuts (especially Brazils and pistachios) – would also fail the 1 ppb test pretty frequently. The economic impact of a 20 ppb policy was not great. The impact of a 1 ppb limit could be very large for these industries. Did it still make scientific sense to pursue an "analytical detection limit" goal, at any cost? Was the scientific evidence about cancer risks at very low intakes that certain?

Here we come to the heart of the problem we shall explore in this book: just how certain is our science on matters such as this? And how should public health officials deal with the uncertainties? We shall be exploring the two responses to the FDA's position that were set out earlier and learn what we can about their relative scientific merits; not specifically in connection with the aflatoxin problem, but in a more general sense. We shall also be illustrating how regulators react to these various scientific responses, and others as well, using some examples where the economic stakes are very high. One would like to believe that the size of the economic stakes would not influence scientific thinking, but it surely influences scientists and policy-makers when they deal with scientific uncertainties.

In the meantime keep in mind that, although considerable progress has been made in reducing aflatoxin exposures, these mold products are still present in some foods, and you have probably ingested a few nanograms (billionths of a gram!) recently. Indeed in many areas of the world, particularly in lesser developed countries, aflatoxin contamination of foods and feeds is widespread. Moreover, the evidence that aflatoxins are a cause of human liver cancer, particularly in individuals affected by hepatitis B virus, has strengthened considerably since the 1970s. The question of the magnitude of the health risk posed by the aflatoxins, and its overall public health significance, remains an important one.

In 1983 a committee of the National Research Council – National Academy of Sciences issued a relatively brief report with red covers entitled: "Risk Assessment in the Federal Government: Managing the Process." The so-called "Red Book" committee had been organized to respond to a request from the US Congress to examine the scientific work of those regulatory agencies that had been given responsibility for enforcing federal laws aimed at guarding the health of people using or otherwise exposed to chemical products of all types, and to chemical contaminants of the environment. These many laws required the regulators to make decisions regarding the introduction of some classes of new chemicals, and to begin setting limits (of the type just described for aflatoxin) on contaminants of air, water, and food, and of workplace environments. The development of knowledge regarding the toxic properties of commercially produced chemicals, and of the by-products of their production, use, and disposal, had accelerated considerably during the decade of the 1970s, as did knowledge of the many chemical by-products of energy production. At the same time, as analytical chemistry improved, knowledge of human exposures to these many products expanded at an even greater rate than did knowledge of their possible adverse health effects. Regulators were activated, and scientists in the various regulatory agencies began turning out what soon came to be called "risk assessments" – documents that attempted to integrate epidemiological and experimental information related to chemical toxicity, with information on human exposures to chemicals, for purposes of evaluating the public health consequences of these exposures. Completion of most risk assessments required the use not only of scientific data, but also the use of various incompletely tested assumptions – about low-dose effects, for example, or about the relevance of experimental animal data to humans. Scientific controversies of many types arose during this time, and scientists in regulatory agencies were often accused of manipulating risk assessment results (by arbitrary adoption of whatever assumptions yielded the preferred result) to satisfy the desires of regulatory decision-makers (to regulate or not, depending upon the political context and climate). Science, it was alleged, both by representatives of regulated industries and by consumer and environmental advocates, was being perverted.

Certain members of the US Congress became convinced that close inspection of "regulatory science" and its uses in the making of regulations was necessary. In fact, some suggested that the risk assessment activities of federal agencies be institutionally separated from the decision-making processes of those agencies; in this way scientists could operate in environments free of political contamination, and simply serve up highly objective risk assessments for use by regulators. As is frequently the case when difficult science policy questions arise, the National Academy of Sciences was asked to offer its opinion. Thus came the Red Book on risk assessment, issued in 1983 after an 18-month study.

The Red Book did much to clear the air, and its influence has been profound. The committee offered clear definitions of risk assessment and of the analytic steps that comprise it, and those definitions and their conceptual underpinnings remain for the most part in place today, not only in the United States but also around the world. The committee clarified the relationships between research, risk assessment, and the set of activities it described as risk management. Risk assessment, the committee insisted, was the critical link between research and decisions about the use of research results for public health protection (through regulation and other means of policy development as well). Most questions regarding risks to health are not answered directly by research scientists. Someone – the risk assessor – needs to evaluate and integrate often diverse and sometimes conflicting sets of research data, and to create a picture of what is known and what is not known about specific health risks, in a form that is (to use an overly fashionable term) transparent, and that is also useful to the risk management context.

Perhaps the most important contribution of the Academy committee came in the area of what its report called "science policy." The term was not used, as it is typically, to describe issues of, for example, the public funding of scientific research, or the priorities given to various research endeavors. In the context of the committee's report, the phrase was used to describe the considerations to be given to the choice of scientific assumptions that are necessary to complete a risk assessment – necessary, because scientific knowledge is incomplete and decisions must be made. Not to act because knowledge is incomplete could clearly jeopardize public health. But to act on the basis of incomplete knowledge could also lead to unnecessary (and often very costly) regulations. Risk assessments conducted using the best available knowledge and, where necessary, assumptions not chosen on an arbitrary, case-by-case basis, but adopted for general application, could bring a greater degree of objectivity to the decision-making process. The Red Book offered a guide to the selection of those general assumptions. The recommendations found in the 1983 report (which is still available from the National Academy and which is a critical resource for acquiring an understanding of the current world of risk analysis policy) provides much of the framework for this book. This book is far more heavily devoted than is the Red Book to the scientific underpinnings for risk assessment, but its discussion of how risk assessment draws upon that science and arrives at results useful for risk management is heavily under the report's influence.

By the way, the committee rejected the suggestion that risk assessment activities be institutionally separated from the risk management activities of regulatory agencies. It recognized the potential problem of the distortion of science, but proposed other, less drastic means to minimize that problem. The committee's thinking on this important matter will emerge in the later chapters of the book.

# 1

# Chemicals and chemical exposures

It is perhaps too obvious to point out that everything we can see, touch, smell, and taste is a chemical or, more likely, a mixture of many different chemicals. In addition, there are many chemical substances in the environment that cannot be detected with the senses, but only indirectly, by the sophisticated instruments scientists have devised to look for them. The number of different chemicals in and on the earth is unknown, but is surely in the many millions. During the past 125 years scientists have been successful in creating hundreds of thousands of compounds that do not occur in nature, and they continue to add to the earth's chemical stores, although most of these synthesized chemicals never leave the research scientists' laboratories.

For both historical and scientific reasons chemists divide up the universe of chemicals into inorganic compounds and organic compounds. The original basis for classifying chemicals as "organic" was the hypothesis, known since the mid nineteenth century to be false, that organic chemicals could be produced only by living organisms. Modern scientists classify chemicals as "organic" if they contain the element carbon.[1] Carbon has the remarkable and almost unique property that its atoms can combine with each other in many different ways, and, together with a few other elements – hydrogen, oxygen, nitrogen, sulfur, chlorine, bromine, fluorine, and a few more – can create a huge number of different molecular arrangements. Each

---

[1] There are a few compounds of carbon that chemists still consider inorganic: these are typically simple molecules such as carbon monoxide ($CO$) and carbon dioxide ($CO_2$) and the mineral limestone, which is calcium carbonate ($CaCO_3$).

such arrangement creates a unique chemical. Several million distinct organic chemicals are known to chemists, and there are many more that will be found to occur naturally or that will be created by laboratory synthesis. All of life – at least life on earth – depends on carbon compounds, and probably could not have evolved if carbon did not have its unique and extraordinary bonding properties, although chemists have verified many thousands of times over that the creation of organic chemicals does not depend on the presence of a living organism.[2]

Everything else is called inorganic. There are 90 elements in addition to carbon in nature (and several more that have been created in laboratories), and the various arrangements and combinations of these elements, some occurring naturally and others resulting from chemical synthesis, make up the remaining molecules in our world and universe. Because these elements do not have the special properties of carbon, the number of different possible combinations of them is smaller than can occur with carbon.

What is meant when a chemical is said to be "known"? Typically this means that chemists have somehow isolated the substance from its source, whether natural or synthetic, have taken it to a relatively high state of purity (by separating it from chemicals that occur with it), have measured or evaluated its physical properties – the temperatures at which it melts, boils, and degrades, the types of solvents in which it dissolves, and so on – and have established its molecular architecture. This last act – determination of chemical architecture, or structure – typically presents the greatest scientific challenge. Understanding chemical structure – the number, type, and arrangement of atoms in a molecule – is important because structure determines how the compound undergoes change to other compounds in chemical reactions, and also how it interacts with biological systems, sometimes to produce beneficial effects (nutrients and medicinal agents) and sometimes to produce harmful effects.

Chemists represent the structures of chemicals using letter symbols to represent atoms (C for carbon, H for hydrogen, O for oxygen), and lines to indicate the chemical bonds that link atoms together (each bond is actually a pair of interacting electrons). The simplest of all

---

[2] Of course the chemists creating organic compounds are living (most of them anyway), but the compounds are created in laboratory flasks without the assistance of living organisms. Such synthesis is clearly different from, for example, the production of colors by flowers and aflatoxins by molds.

organic compounds, the naturally occurring gas methane (marsh gas), has the structure represented below, in two dimensions.

```
        H
        |
   H —  C  — H
        |
        H
```

Methane

Molecules of methane are actually three-dimensional, with the carbon atom at the center of a tetrahedron and a hydrogen atom at each of the four corners. When we refer to the chemical methane, we refer to collections of huge numbers of these specific molecules. An ounce of methane contains about $12 \times 10^{23}$ (12 followed by 23 zeros) of these molecules.

An interesting, important, and common phenomenon in organic chemistry is that of structural isomerism. Consider a molecule having two carbon atoms, four hydrogen atoms, and two chlorine atoms $(C_2H_4Cl_2)$. These atoms are capable of binding to each other in two different ways as shown.

```
     H   H                    H   H
     |   |                    |   |
Cl — C — C — Cl          H — C — C — Cl
     |   |                    |   |
     H   H                    H   Cl
```

1,2-Dichloroethane          1,1-Dichloroethane

More ways are not possible because of limitations on the number of bonds each type of atom can carry (carbon has a limit of four, hydrogen and chlorine have a limit of one each). But the important lesson here is that the two molecules shown (1,1-dichloroethane and 1,2-dichloroethane) are different chemicals. They have identical numbers of C, H, and Cl atoms ("isomer" means "same weight") but different chemical structures.

These two chemicals have different physical and chemical properties, and even produce different forms of toxicity at different levels of exposure. The way chemicals interact with biological systems to produce damage depends greatly upon details of molecular structure, although our understanding of how structures affect those interactions is relatively poor.

The structures of inorganic compounds are represented by the same types of conventions as shown for the three organic compounds depicted above, but there are some important differences in the nature of the chemical bond that links the atoms. These differences need not concern us here, but will have to be mentioned later in the discussion. Toxicologists refer to many of the environmentally important inorganic chemicals simply according to the name of the particular portion of the compound that produces health damage. They refer, for example, to the toxicity of lead, mercury, or cadmium, without reference to the fact that these metals are actually components of certain compounds that contain other elements as well. Sometimes the particular form of the lead or mercury is important toxicologically, but often it is not. Toxicologists tend to simplify the chemistry when dealing with metals such as these. Metals can also exist in different states of oxidation, and the toxicity of a metal can vary with oxidation state. Chromium in the so-called (+3) oxidation state is an essential nutrient and shows little toxicity even at high exposures. In a higher oxidation state (+6), chromium is a respiratory carcinogen when inhaled.

## Naturally occurring chemicals

Living organisms contain or produce organic chemicals, by the millions. One of the most abundant organic chemicals on earth is cellulose; a giant molecule containing thousands of atoms of C, H, and O. Cellulose is produced by all plants and is the essential structural component of them. Chemically, cellulose is a carbohydrate (one that is not digestible by humans), a group which, together with proteins, fats, and nucleic acids are the primary components of life. But as mentioned in the Prologue in connection with the chemistry of molds, living organisms also produce huge numbers of other types of organic molecules. The colors of plants and animals, and their odors and tastes are due to the presence of organic chemicals. The numbers and structural varieties of naturally occurring organic chemicals are staggering.

Other important natural sources of organic chemicals are the so-called fossil fuels – natural gas, petroleum, and coal – all deposited in the earth from the decay of plant and animal remains, and containing thousands of degradation products. Most of these are simple compounds containing only carbon and hydrogen (technically and even reasonably known as hydrocarbons). Natural gas is relatively simple in composition and is mostly made up of gases such as methane (marsh gas, already described above). The organic chemical industry depends upon these and just a few other natural products for everything it manufactures; the fraction of fossil fuels not used directly for energy generation is used as "feedstock" for the chemical industry. There are, or course, inorganic chemicals present in living organisms, many essential to life – the minerals. But the principal natural source of inorganic chemicals is the non-living part of the earth that humans have learned how to mine.

It may be a surprise to some that the largest numbers of chemicals to which humans are regularly and directly exposed are the natural components of the plants and animals we consume as foods. In terms of both numbers and structural variations, no other chemical source matches food. We have no firm estimate of the number of such chemicals we are exposed to through food, but it is surely immense. A cup of coffee contains, for example, well over 200 different organic chemicals – natural components of the coffee bean that are extracted into water. Some impart color, some taste, some aroma, others none of the above. The simple potato has about 100 different natural components (some quite toxic, as shall be seen), and to make matters more interesting and confusing, some of the chemicals found in the potato and the amounts present vary among different varieties and even different conditions of cultivation and storage! Products of fermentation – cheeses, wines – contain huge numbers of chemicals not present in milk and grapes. Herbs and spices consist of thousands of organic chemicals, some quite unusual in structure. The issue of naturally occurring food constituents will come up several times in this book.

## Synthetic chemicals

The decade of the 1850s is noted by historians of science as significant because it saw the publication of Darwin's *Origin of Species* (1859), a work that has had profound influences on contemporary society. But other scientific events occurred at about the same time, which, I would argue if I were a historian, had nearly equal significance for our time. They allowed the development of organic chemical science, and so greatly increased our understanding of chemical behavior that they spawned the age of chemical synthesis. Chemical synthesis is the science (one might say art) of building chemicals of specified structure from simpler and readily available chemicals, usually petroleum or coal products, and other natural chemicals. Sometimes chemists engage in synthesis for rather obscure purposes, related to gaining some understanding of fundamental chemical principles. More often, they are interested in creating molecules that possess useful properties.

Chemists and their historical predecessors have for dozens of centuries manipulated natural products to make useful materials, but for most of this time they had little understanding of what they were doing. It wasn't until the "structural theory of organic chemistry" began to solidify during the third quarter of the nineteenth century that chemists could possibly understand the molecular changes underlying such ancient arts as fermentation, dyeing, and soap making. The art of chemical purification had been well developed by the mid nineteenth century, but again chemists couldn't say much about the properties of the substances they had purified (mostly acids, alcohols, and aromatic chemicals from plants and animals) until structural theory came along. But once chemists grasped structural theory it became possible to manipulate chemicals in a systematic way so that certain molecular arrangements could be transformed in predictable ways to other, desired arrangements. The chemists who developed the structural theory of organic compounds, and those who applied it to the synthesis of substances that were not products of nature but totally new to the world, were instigators of a mammoth industrial revolution, one that has given us an extraordinary variety of beneficial materials.

Through the efforts of many chemical pioneers, mostly European, organic chemical science began to take on its contemporary shape during the first half of the nineteenth century. It was not until 1858, however, that Friedrich August Kekulé von Stradonitz, a student of architecture who had become captivated by chemical science and who held a position at the University of Heidelberg, and Archibald Scott Couper, a Scotsman then at the Sorbonne, independently introduced the so-called "rules of valency" applicable to compounds of carbon. This work unified thinking about the structural characteristics of organic chemicals, because it allowed chemists for the first time to explain hundreds of early observations on the chemical behavior of organic chemicals, and, as noted, structural theory also set the stage for the rapid development of the chemical industry. It is a distortion to say that Kekulé and Couper single-handedly formulated the structural theory of organic chemistry – they built upon and synthesized the earlier and quite extraordinary work of giants such as Edward Frankland (1825–99) at Manchester, Justus von Liebig (1803–73) at Geissen, Joens J. F. von Berzelius (1779–1848) at Stockholm, Friederick Wohler

(1800–82) at Göttingen, Marcellin Bertholet (1827–1907) at Paris, and several others as well. The publications by Kekulé and Couper in 1858 nevertheless clearly unleashed powerful new forces in chemical science and their societal repercussions have been profound, in practical ways perhaps more so than those produced by the Darwinian revolution.[3] Most history books do not seem to capture the work of these great scientists, nor have historians extensively explored their legacy for the modern world. Part of that legacy – the possible adverse public health consequences of their work – is the topic of this book. As that legacy is explored we should probably keep in mind the enormous benefits that are also part of that legacy.

The development of the chemical industry did not, of course, spring wholly from the work of theoreticians such as Kekulé. William Henry Perkin (1838–1907), working at the age of 18 in the laboratory of August Wilhelm von Hofmann at the Royal College of Science in London, had been put to work on the synthesis of the drug quinine from aniline, the latter a coal-tar product that had been isolated by Hofmann. Perkin failed to synthesize quinine, but as a result of his

---

[3] Kekulé's other major contribution was his hypothesis, later shown to be correct in essentials though not in details, regarding the structure of the petroleum hydrocarbon known as benzene. This important chemical has the molecular formula $C_6H_6$, and it was Kekulé who first recognized, in 1865, that the six carbon atoms link to each other to form a ring:

Two presentations of the structure of benzene

Kekulé reported his discovery as follows:

I was sitting, writing at my text-book; but the work did not progress, my thoughts were elsewhere. I turned my chair to the fire and dozed. Again the atoms were gambolling before my eyes. This time the smaller groups kept modestly in the background. My mental eye, rendered more acute by repeated visions of the kind, could now distinguish larger structures, of manifold conformation: long rows, sometimes more closely fitted together; all twining and twisting in snake-like motion. But look! What was that? One of the snakes had seized hold of its own tail, and the form whirled mockingly before my eyes. As if by a flash of lightning I awoke; and this time also I spent the rest of the night in working out the consequences of the hypothesis. Later he adds "Let us learn to dream, gentlemen, then perhaps we shall find the truth . . . but let us beware of publishing our dreams before they have been put to the proof by the waking understanding."

A snake biting its own tail seems a far remove from structural theory, but how many scientific advances arise in just such ways?

efforts, his flasks and just about everything else in his laboratory consistently ended up stained with a purplish substance. Perkin called this substance "aniline purple," and when the French found it an excellent material for dyeing fabrics they named it mauve. The color became an immensely successful commercial product, and its widespread use created the "mauve decade" of the nineteenth century. At the time of Perkin's discovery (which was by no means the result of a planned, systematic synthesis based on an understanding of structural theory) dyes were derived from natural sources. It wasn't long after the young scientist's discovery that the synthetic dye industry was born; it was the first major industry based upon the science of organic synthesis, and it flourished, especially in Germany.

These two events, the solidification of structural theory and the discovery of a useful and commercially viable synthetic chemical, gave birth to the modern chemical world – the world that is much of the subject of this book. Industrial applications have led to the introduction into commerce of tens of thousands of organic chemicals, unknown to the world prior to the 1870s, and the sciences of toxicology and epidemiology grew in response to the need to understand how these new substances might affect the health of workers involved in their production and use, and of the rest of the population that might be exposed to them. It is of interest, then, to sketch a portrait of the organic chemical industry.

## Industrial organic chemistry

Laboratory synthesis of organic chemicals proceeded at an astounding pace following the introduction and success of structural theory, and further impetus was provided by the economic success of the synthetic dye industry. Chemists have learned thousands of different ways to manipulate groups of atoms in organic compounds to create new molecular arrangements, and have also found how to develop sequences of individual chemical transformations that could lead to molecules of desired structural arrangement. They have learned how to lay out on paper a "blueprint" for creating a molecule of specific structure and how to achieve that plan in the laboratory (although, of course, many plans fail to be achieved). If the synthesis is successful and the product useful, chemical engineers are called in to move the laboratory synthesis to the industrial production stage. Tens of thousands of synthetic organic chemicals and naturally occurring chemicals are in commercial production.

In the late nineteenth century and up to World War II coal was the major "starting material" for the organic chemical industry. When coal is heated in the absence of oxygen, coke and volatile by-products called coal tars are created. All sorts of organic chemicals can be isolated from coal tar – benzene, toluene, xylenes, ethylbenzene, naphthalene, creosotes, and many others (including Hofmann and Perkin's aniline). The organic chemical industry also draws upon other natural products, such as animal fats and vegetable oils, and wood by-products.

The move to petroleum as a raw materials source for the organic chemical industry began during the 1940s. Petrochemicals, as they are called, are now used to create thousands of useful industrial chemicals. The rate of commercial introduction of new chemicals shot up rapidly after World War II.

Some of these chemicals are used primarily as *solvents* for one purpose or another. Many of the hydrocarbons found in petroleum, such as gasoline and kerosene, and the individual chemicals that make up these two mixtures, are useful as non-aqueous solvents, capable of dissolving substances not readily soluble in water. Hydrocarbons are highly flammable, however, and chemists found that conversion of hydrocarbons to chlorine-containing substances, using reactions in which hydrogen atoms were replaced by chlorine (below), create solvents still useful for dissolving substances not soluble in water, but having reduced flammability. In the 1940s and 1950s, solvents such as these came into very wide use in the organic chemical industry and in many other industrial settings where they were needed (for degreasing of oily machinery parts, for example, or for "dry" cleaning of clothing). As is now commonly acknowledged, we get no "free lunch." The so-called "chlorinated hydrocarbon" solvents tend to be more toxic and more persistent in the environment than are the petroleum hydrocarbons. Indeed two very widely used solvents, perchloroethylene and trichloroethylene, are among the most common contaminants of ground water. Many commercial uses of trichloroethylene have vanished, but "perc" is still used in a number of applications including dry cleaning. Reduced risk of fire and explosion have thus brought increased risk of environmental harm. At the time industrial decisions were taken to move to the less flammable solvents, these kinds of trade-off were little discussed; corporations are still learning how to balance risks and benefits, but now they are at least aware that decisions of this type should not proceed without appreciable understanding of the environmental consequences.

Chloroform      Perchloroethylene      Trichloroethylene

Among the thousands of other products produced by the organic chemical industry and by related industries are included, in no particular order, medicines (most of which are organic chemicals of considerable complexity), dyes, agricultural chemicals including substances used to eliminate pests (insecticides, fungicides, herbicides, rodenticides and other "cides"), soaps and detergents, synthetic fibers and rubbers, paper chemicals, plastics and resins of great variety, adhesives, substances used in the processing, preservation, and treatment of foods (food additives), additives for drinking water, refrigerants, explosives, cleaning and polishing materials, cosmetics, textile chemicals.

People can be exposed to a greater or lesser degree to most of these chemicals, indeed human exposures for many, such as medicines, are intended; and some people are exposed to other chemicals used in or resulting from production – the starting materials, the so-called "intermediates" that arise during synthesis but which are not in the final products, and by-products of their production, including solvents and contaminants.


## Inorganic chemicals and their production

The history of human efforts to tap the inorganic earth for useful materials is complex and involves a blend of chemical, mining, and materials technologies. Here we include everything from the various silicaceous materials derived from stone – glasses, ceramics, clays, asbestos – to the vast number of metals derived from ores that have been mined and processed – iron, copper, nickel, cadmium, molybdenum, mercury, lead, silver, gold, platinum, tin, aluminum, uranium, cobalt, chromium, germanium, iridium, cerium, palladium, manganese, zinc, etc. Other, non-metallic materials such as chlorine and bromine, salt (sodium chloride), limestone (calcium carbonate), sulfuric acid, and phosphates, and various compounds of the metals, have hundreds of different uses, as strictly industrial chemicals and as products consumers use directly. These inorganic substances reach, enter, and move about our environment, and we come into contact with them, sometimes intentionally, sometimes inadvertently. As with the organic chemicals, this book is about the potential health consequences of these contacts and exposures.

## The newest thing – products of nanotechnology

A potentially vast industry is now in early bloom, one based on a technology that has been under development for several decades. It pertains to the science of the very, very small. Chemists, physicists and engineers have learned how to produce chemicals in solid particle form having at least one, and sometimes two or three dimensions, at the so-called "nano" scale. A nanometer is one-billionth of one meter, and this is the scale that describes the size of some individual cells, viruses, cell walls, and individual components of cells. Many of these "nanoengineered" particles take on properties remarkably different from those of the very same chemical when it is produced, as normal, having much larger dimensions. Some engineered nanomaterials are already in commercial production and it appears many more are coming. Potential uses are many and diverse. We shall see in Chapter 9 that the very small size of these particles may give rise to unusual forms of toxicity, and that managing their health and environmental risks may present significant challenges.

## Human exposure to chemicals

A necessary, indeed critical, element in understanding whether a chemical's toxic properties will be expressed is the size of the dose incurred by individuals who are exposed. How exposure leads to dose is the subject of the next chapter. Here we set the stage for the next chapter by describing the many ways in which people can become exposed to all the categories of chemicals we have described.

Our survival and that of all plants and animals on our planet depends upon chemicals: water, the nutrients in our foods, oxygen in our air, and if you are a plant, carbon dioxide as well. Oxygen, carbon dioxide, and water are simple inorganic chemicals, and the major nutrients – proteins, carbohydrates, fats, and vitamins – are organic compounds of considerable complexity. Certain inorganic minerals – calcium, zinc, iron, copper, sodium, potassium, and a few others – are also essential to life. Living organisms can be envisioned as complex, highly organized collections of chemicals that absorb other chemicals from the environment and process them in ways that generate and store the energy necessary for survival of the individual and the species, primarily by the making and breaking of chemical bonds. It is a process of staggering complexity, and beauty, and reveals in intricate detail the extraordinary interdependence of living organisms and their environments that has resulted from the processes of evolution.

To maintain health, human beings and all other life forms must ingest, inhale, and otherwise absorb the essential environmental chemicals only within certain limits. If the amounts we take into ourselves fall below a certain level, we may suffer malnutrition or dehydration, or suffer the effects of oxygen deprivation. If we take in too much, we become obese, develop certain forms of cancer, heart disease, and many other diseases as well. We can even poison ourselves by consuming large amounts of essential vitamins and minerals. Living organisms can certainly tolerate a fairly wide range of intake of these essential chemicals, but there are limits for all of them beyond which we get into trouble. Scientists are still learning a great deal about what those limits are, and as more is learned we shall no doubt be able to put this knowledge to use to reduce the risk of a number of important diseases – perhaps the most burdensome diseases we experience – and to enhance the health and well-being of all life.

When we take a drink of water we expose ourselves not only to molecules of $H_2O$, but to other chemicals as well. Depending upon the source of the water, we are typically ingesting a variety of minerals, some of which are essential to health, but many of which have no known role in preserving our health. We are also consuming some organic chemicals that have migrated from plants or soil microorganisms. These minerals and organic chemicals are naturally occurring, but in many areas, drinking water will also contain certain minerals at concentrations in excess of natural levels because of some type of human activity – mining, manufacturing, agriculture – and synthetic organic chemicals that have somehow escaped into the environment. We also intentionally add some chemicals to water to achieve certain technical effects, including the very important effect of disinfection.

The situation is the same with the air we breathe. We need oxygen, and so cannot avoid nitrogen, carbon dioxide and several other naturally occurring gases. We are also inhaling with every breath a variety of naturally occurring and industrial chemicals that are either gases or are liquids volatile enough to enter the gaseous state. We have learned in the past decade much about the public health importance of very fine particles present in ambient air, particles of diverse chemical composition produced by fires, the internal combustion engine, power plants, and other industrial and natural processes. The dimensions of these particles extend from the so-called ultrafine range (similar to the dimensions of engineered nanoparticles) to the easily visible macro scale, but it is the fine and ultrafine range that presents the most significant public health problem.

The plants and animals we have chosen to use as foods naturally contain, as we have already noted, thousands of chemicals that have no nutritional role, and when we eat to acquire the nutritionally essential chemicals we are automatically exposed to this huge, mostly organic, chemical reservoir. Of course, human beings have always manipulated foods to preserve them or to make them more palatable. Processes of food preservation, such as smoking, the numerous ways we have to cook and otherwise prepare food for consumption, and the age-old methods of fermentation used to make bread, alcoholic beverages, cheeses and other foods, cause many complex chemical changes to take place, and so result in the introduction of uncounted numbers of compounds that are not present in the raw agricultural products.

Human beings have for many centuries been quite good at manipulating the genetic material of food plants and animals to produce varieties and hybrids with improved characteristics of one type or another. For most of the time breeders had no idea what they were doing, because little was known about genetics and the consequences of manipulating genetic material. Now we know that when we create a new tomato hybrid we are producing changes in the chemical composition of the fruit; from the chemist's point of view, a tomato is not a tomato is not a tomato. People eating the variety of tomatoes available 50 years ago did not ingest exactly the same collection of compounds found in varieties available nowadays.

Spices and herbs contain no nutritionally essential chemicals of consequence, but they do contain hundreds of organic compounds which impart flavors, aromas, and colors. Here, people deliberately expose themselves to an abundance of unusual chemicals largely for aesthetic reasons.

Like air and water, foods may also become contaminated with industrial chemicals, and certain unwanted, naturally occurring substances such as aflatoxins. For centuries, certain chemicals have been added to food to achieve a variety of technical effects.

We also paint our bodies with all sorts of colors, and splash perfumes and other cosmetics on our skin. We wash ourselves and our environments with chemicals. We take medicines, many of which are exceedingly complex organic molecules. We use chemicals to rid ourselves and our food of pests. Our bodies contact materials – chemicals – we use for clothing and to color clothing. Most of us are exposed to chemicals on the job and through the hundreds of products we use in the house and for recreation. Fuels and products resulting from their combustion and materials used for buildings add to the burden. In the United States 30 million people can't seem to avoid the several thousand chemicals they inhale after every puff of a cigarette. Even "side-stream" and exhaled smoke contain hundreds of chemicals.

Food, air, water, consumer products, cosmetics, pesticides, medicines, building materials, clothing, fuels, tobacco products, materials encountered on the job, and unwanted contaminants of all of these – these are the principal sources of the thousands, perhaps hundreds of thousands of known and unknown chemicals, natural and industrial, that people ingest, inhale, absorb through their skin, and (for medicines) take into their bodies in other ways. We should also mention unusual, but not uncommon, forms of exposure: venoms and other substances from animals that may bite or sting us and plants with which we may come into contact, and soils and dusts we inadvertently or intentionally ingest or inhale. Indeed, recent studies reveal that common house dusts may contain dozens of different commercial chemicals. Chemical accidents resulting in the release of dangerous products are not rare events. Chemical compounds having extremely high toxicity are weapons of the modern terrorist.

As with the very small fraction of these chemicals that are essential to our health, it appears there are ranges of exposure to all these chemicals that, while probably not beneficial to health, are probably without detrimental effect. And, for all of them, both the industrial ones and those of strictly natural origin, there are ranges of exposure that can put our health into jeopardy. These facts bring out one of the most important concepts in toxicology: all chemicals are *toxic* under some conditions of exposure. The toxicologist attempts to understand what those conditions are. Once they are understood, measures can be taken to limit human exposure so that toxicity can be avoided. Of course the latter can be achieved only if societies have appropriate control and enforcement mechanisms in place.

In the next chapter, and for a large part of the book, toxicity will be the main topic. For now we need only note that by *toxicity* we mean the production of any type of damage, permanent or impermanent, to the structure or functioning of any part of the body.

## Environmental media

Chemicals reach us through various *media*. By media we mean the vehicles that carry the chemical and that get it into contact with the body. Thus, food, beverages, air, water, and soils and dusts are the principal environmental media through which chemical exposures take place. Direct contact with the chemical, as with cosmetics applied to the skin, or household products accidentally splashed into the eye, may also occur, in which case the cosmetic or household product may be said to be the medium through which exposure occurs. Exposures to medicines occur by ingestion of tablets containing them, by injection, and by other means. Sometimes workers come into direct contact with the substances they are using.

## Exposure pathway

It is often of interest to understand the *pathway* a chemical uses to reach the medium that ultimately creates the exposure. This typically takes the form of a description of the movement of a chemical through various environmental media. Some pathways are short and simple: aflatoxin, for example, contaminates a moldy food and we consume that food; there is only one medium (the food) through which the aflatoxin moves to reach us. Most pathways are somewhat more complex.

Lead added to gasoline (medium 1) is emitted to the air (medium 2) when gasoline is burned. (Although the use of leaded gasoline has been eliminated in much of the world, there is still widespread use, and the legacy of past uses is still with us.) Some of the airborne lead deposits in soil (medium 3) which is used for growing corn. Some of the lead in soil dissolves in water (medium 4) and moves through the roots of the corn plant, accumulating in the kernels of corn (medium 5). The corn is fed to dairy cattle and some of the lead is excreted in milk (medium 6). Milk is the medium that creates human exposure to the lead. The lead has passed through six media to reach a human being. To make matters more complex, note that people may be exposed to lead at several other points along the pathway, for example by breathing the air (medium 2) or coming into contact with the soil (medium 3).

Knowledge of exposure pathways is a critical part of the analysis needed to piece together the human exposure pattern.

## Exposure routes

The *route* of exposure refers to the way the chemical moves from the exposure medium onto or into the body. For chemicals in the environment the three major routes are *ingestion* (the oral route), *inhalation*, and *skin contact* (or *dermal* contact). Medicines get onto or into the body in these three ways, but in several other ways as well, for example by injection under the skin or directly into the bloodstream, or by application to the eye.

In many cases, a given medium results in only one route of exposure. If diet is the medium, then ingestion will be the exposure route. If air is the medium, the chemical enters the body by inhalation. Immediately, however, we can think of ways this simple rule does not hold. Suppose, for example, that a chemical is contained in very small particles of dust that are suspended in air. The air is inhaled and the dust particles containing the chemical enter the airways and the lungs. But some of these dust particles are trapped before they get to the lungs and others are raised from the lungs by a physiological process to be discussed later. These particles can be collected in the mouth and swallowed. So here is an example of a single medium (air) giving rise to two exposure routes (inhalation and ingestion). These types of possibilities need to be considered when exposures are being evaluated.

Exposure media, pathways, and routes are important determinants of dose. We shall see how this is so in the next chapter.

## Chemicals and our laws

Because all of these various types of chemical exposures can, under the right conditions, be damaging to health, our legislators have over the past 50–60 years developed extensive sets of laws that impose requirements upon manufacturers, distributors, and commercial users of chemicals, and upon those who dispose of chemical wastes. Several of these laws also apply to other sources of chemical exposure, such as power plants and moving vehicles. The laws tend to divide up according to environmental media, source of chemical, or type of population exposed. Thus, in the United States, there are federal laws specific to air and water, there is a law covering occupational exposures, another covering pesticidal chemicals, a couple pertaining to hazardous wastes, and one covering industrial chemicals not controlled under other laws. There is the Food, Drug, and Cosmetics Act (the first version of which appeared in 1906), and laws pertaining to other types of consumer products. There is even a different food law pertaining specifically to meat and poultry products. Each of these laws has its own history and unique set of requirements. In the United States they are administered by the Environmental Protection Agency (EPA), the Occupational Safety and Health Administration (OSHA), the Food and Drug Administration (FDA), the Consumer Product Safety Commission (CPSC), and the Food Safety and Inspection Service (FSIS) of the Department of Agriculture. Other laws having to do with chemical exposure are administered by the Department of Transportation (DOT), the Agency for Toxic Substances and Disease Registry (ATSDR), the Centers for Disease Control (CDC), and the National Institute for Occupational Safety and Health (NIOSH). The latter three agencies do not have the same authority as the regulatory agencies, but have critical roles of other types. Similar legal and institutional arrangements are established in the European Union, Japan, and the rest of the technologically advanced world, and exist in various stages of development in other countries.

A near-lifetime of experience is required to grasp these many laws and the thousands of regulations that have flowed from them. But with respect to chemical exposure, all direct the administrative agencies to impose requirements that, in somewhat different ways, are directed at protecting human health from levels of exposure that could be harmful. Risk assessment, the principal subject of this book, is now the procedure commonly used to understand the risks the exposures might entail. The laws impose various types of requirements upon exposures found to create excessive risks to human health (and, although it is outside the subject of this book, to the health of the environment). After completing our tour of toxicology and risk assessment we shall return, in Chapter 11, to the subject of risk management, and discuss these various requirements.

So, with this larger social context in mind, we return to the technical discussion, and move from exposure to what is a principal determinant of risk: the dose.

# 2

# From exposure to dose

To determine whether and to what extent humans may be harmed (suffer toxicity) from a chemical exposure, it is necessary to know the *dose* created by the exposure. The concept of dose is so important that it needs to be treated in detail.

Exposure, as we have seen, refers to an individual's contact with an environmental medium containing a chemical or, in some cases, with the chemical itself. The amount of chemical that enters the body as a result of the exposure is called the dose.

The magnitude of the dose is a function of the amount of chemical in the medium of contact, the rate of contact with the medium, the route of exposure, and other factors as well. Experts in *exposure analysis* use various means to estimate the dose incurred by individuals exposed to chemicals. Exposure analysis is one of the critical steps in toxicological risk assessment.

## Common expressions of dose

Everyone is generally familiar with the term dose, or dosage, as it is used to describe the use of medicines. A single tablet of regular strength aspirin typically contains 325 milligrams (mg) of the drug. An adult takes four tablets in one day, by mouth. The total *weight* of aspirin ingested on that day is 1300 mg, or 1.3 grams (1 mg = 0.001 g). But weight is not dose. The *dose* of aspirin varies with the weight of the individual consuming it, and typically dose is expressed as the amount taken into the body divided by the weight of the person, expressed in kilograms (1 kg = 1000 g). The aspirin *dose* for a 65 kg adult (about 145 lbs) is thus 1300 mg/65 kg = 20 mg/kg body weight (b.w.). The time over which the drug was taken is also important in describing dose. Our adult took four tablets in one day, and the day is the usual time unit of interest. So a more complete description of the aspirin dose in this case is given by 20 mg/(kg b.w. day). The typical *dose units* are thus milligrams of chemical per kilogram of body weight per day (mg/(kg b.w. day)).

Note – and this is quite important – that if a 20 kg child (about 44 lbs) were to take the same four tablets on one day, the child's dose would be more than three times that of the adult, as follows: 1300 mg aspirin/20 kg b.w. = 65 mg/(kg b.w. day). For the same intake of aspirin (1300 mg), the smaller person receives the greater dose.

Calculating doses for environmental chemicals is not much more complex. Suppose a local ground water supply has become contaminated with the widely used degreasing solvent, trichloroethylene (TCE). Environmental Protection Agency scientists have measured the extent of TCE contamination and found that the water contains 2 micrograms of TCE in each liter of water (how they know this will be discussed a little later). People living above the water have sunk wells, and adults are drinking an average of two liters of the water each day and their children are consuming one liter.

First we calculate the *weight* of TCE (in mg) getting into their bodies.

*Adults*
Consuming 2 liters per day of water containing 2 micrograms of TCE in each liter leads to ingestion of 4 micrograms of TCE each day.
A microgram is 0.001 mg.
Therefore, weight of TCE taken into the body is 0.004 mg/day.

*Children*
Consuming 1 liter per day of the same water leads to an intake of 0.002 mg/day.

If the adults weigh 80 kg and the children weigh 10 kg, then their respective daily *doses* are:

*Adults*
0.004 mg per day/80 kg b.w. = 0.00005 mg/(kg b.w. day).

*Children*
0.002 mg per day/10 kg b.w. = 0.0002 mg/(kg b.w. day).

The child again receives a higher dose (4 times) than the adult consuming the same water, even though the child consumes less water each day; the difference relates to the smaller body weight of the child.

Ultimately it must be asked whether a health risk exists at these doses, and a second important factor in making this determination is the number of days over which the dose continues. *Duration* of exposure as well as the dose received from it thus needs to be included in the equation.

To be sure of our terms: dose and its duration are the critical determinants of the potential for toxicity. Exposure creates the dose. In our example, people are *exposed* to TCE through the *medium* of their drinking water, and receive a *dose* of TCE by the oral *route*.

Calculating (or, more correctly, estimating) dose requires knowledge of the weight of the chemical getting into the body by each route. As in the account just given of exposure to a chemical in drinking water, the amount of chemical in a specified volume of water (mg/l) times the amount of that water consumed by a person each day (l/day) gave the weight of chemical taken into the body each day. If instead of in drinking water, the chemical is in air, then the required data are weight of chemical for a given volume of air (usually mg per cubic meter, $m^3$) and the volume of air a person breathes each exposure period ($m^3$ per day). Suppose the air in a gasoline station contains 2 mg carbon monoxide per cubic meter and a worker breathes that air for an 8 hour work day. Typically, an adult engaged in a moderately high level of activity will breathe in 10 $m^3$ of air in 8 hours. Thus 2 mg/$m^3$ × 10 $m^3$/8 hrs = 20 mg/8 hrs. If the worker does not inhale carbon monoxide for the rest of the day away from the station, then the daily intake is 20 mg. In fact, the worker will probably also inhale smaller amounts for the rest of the day from other sources and thus receive a somewhat higher total daily dose. The dose may climb significantly if the worker happens to be a smoker.

Food intakes are estimated in the same way. The amount of chemical per unit weight of food is multiplied by the daily intake of that food item to obtain the weight of chemical ingested each day. Dose is then calculated by dividing the weight of the chemical by the body weight of the exposed person. Estimating food intake is not quite as straightforward as estimating air volumes inhaled or water volumes consumed. People's intakes of different foods vary greatly. If we need to estimate the amount of a pesticide someone might receive from residues on treated apples, it is critical to understand whether we are interested in the average eater of apples (40 g per day, or about

1.3 ounces) or above-average eaters. Five percent of the US population consumes, for example, more than 140 g of apples per day (about one-third of a pound), and these people would obviously be exposed to a greater dose of the pesticide than the average eater, assuming all apples had the same amount of pesticide residue (which is unlikely). Toxicologists are generally concerned to ensure that risks to individuals exposed to higher than average amounts of chemicals are avoided, although highly eccentric eating patterns are very difficult to take into account.[1]

Information on food consumption rates and their distribution in the population is available, but it is far from perfect. Because people's food consumption habits change over time, and because so much eating now takes place in restaurants and from foods prepared outside the home, it is becoming increasingly difficult to acquire reliable data on this subject. As with so many other aspects of the science of toxicology and risk assessment, the needed data are either lacking or highly uncertain. Scientists in this field are under increasing pressure to supply firm answers without the benefit of firm data.

Note that in these several examples certain kinds of assumption are used to estimate intakes. In the TCE examples all adults were assumed to consume 2 liters of water each day and were also assumed to weigh 80 kg. Obviously in any population exposed to the contaminated water, it is unlikely that these two assumptions apply with high accuracy to any actual individuals. In fact the assumptions may be quite inaccurate for some individuals, even while they might be reasonably representative, on average, for most. It is in fact not possible to conduct risk assessments without the use of assumptions such as these, and so the "individuals" that are the subjects of typical risk assessments might be described as "generic" rather than actual. As will become clear in the later chapters on risk assessment, this type of generic evaluation is appropriate and useful for the purposes of public health protection.

## Concentrations

Several references have been made to the amount of chemical in a given amount of medium. The technical term that describes this relationship is *concentration*, sometimes referred to as *level*. The concentrations of

---

[1] In a survey conducted at the time of the saccharin scare in 1976, a tiny portion of the US population was found to consume more than 36 cans of diet soda each day!

chemicals are typically expressed in different, though similar, units, according to the medium:

| Medium | Concentration units |
|--------|---------------------|
| water  | milligram chemical/liter water (mg/l) |
| food   | milligram chemical/kilogram food (mg/kg) |
| air    | milligram chemical/cubic meter air (mg/m$^3$) |

A more confusing but widely used unit is that of "parts per." Because such units are often used by TV and the press, and in other public presentations of information on chemicals in the environment, they shall be described here.

For water, "parts per" refers to parts of chemical per so many parts of water. If there are 10 milligrams of TCE in one liter of water, then the concentration is said to be 10 parts-per-million (ppm). How does this come about? First, it must be recognized that one liter of water weighs exactly one kilogram. One milligram is one-thousandth of one gram. One kilogram is 1000 grams. Thus, a milligram is *one-millionth* of one kilogram. So, one milligram TCE per kilogram water is also one part per million. We then see that 10 mg/l is equivalent to 10 ppm.

If instead of mg of the chemical, it is found to be present at a concentration of micrograms/liter, then the equivalent units are parts-per-billion (ppb), because a microgram is one-billionth of a kilogram. Because these units will come up throughout the book, a reference table on them will come in handy:

| Some measures of weight and volume: Metric system | |
|---------------------------------------------------|---|
| **Weight** | |
| kilogram (kg) (1000g) | (approx 2.2 lbs) |
| gram (g) | (approx 1/30 oz) 0.001 kg |
| milligram (mg) | 0.001 g |
| microgram (μg) | 0.000 001 g |
| nanogram (ng) | 0.000 000 001 g |
| picogram (pg) | 0.000 000 000 001 g |
| **Volume** | |
| liter (l) | (approx 0.9 US quart) |
| milliliter (ml) | 0.001 l |
| cubic meter (m$^3$) | 1000 l |

The concentrations of chemicals in environmental media can be known in two major ways. They can be known either because the chemical has been added to known amounts of a medium in known quantities, or because the concentrations have been measured using any of several technologies developed and applied by analytical chemists.

Certain substances are deliberately added to food to achieve desired technical effects: to preserve, to color, to stabilize, to emulsify, to sweeten, and so on. Such substances are added under carefully controlled conditions so that the concentrations of additives in food are generally known with a high degree of accuracy. Pharmaceuticals are manufactured very carefully to achieve specified concentrations of these substances in whatever delivery vehicle is used.

The situation is not so simple with most agents in the environment, and concentrations have to be measured. All such measurements have limitations.

Suppose a pesticide is applied to tomatoes to control fungal invasion. To gain approval for the use of such a pesticide in the United States, the manufacturer would have had to conduct studies and submit results to the EPA regarding the amount of pesticide (in this case a fungicide) to be applied to insure effective fungal control, and the amount of the fungicide that would remain on the tomatoes at harvest time. If the EPA judged such pesticide residue concentrations to be safe (actually, judgment is made on the dose a person would receive from consuming the tomatoes with the specified residue concentration), the manufacturer could receive approval to market the fungicide. This approval is called a pesticide registration. The marketed product would be required to carry a label specifying the approved rate of application, so that the use of the fungicide would result in residue concentrations no greater than those considered by the EPA to be safe.

Knowledge of the concentrations of the fungicide in tomatoes and tomato products depends in this case on measurement. In theory, one might estimate the amount of fungicide on a tomato by calculation, based on the rate of application to the crop. But this is for several reasons an extremely uncertain calculation. To calculate accurately it is necessary to know how much of the applied chemical is actually on the tomatoes as against the remaining parts of the plant. It is also necessary to understand how much might be lost by certain physical processes – rain, wind – as the tomato grows. The rate of chemical degradation of the fungicide has to be understood. The extent to which the fungicide becomes concentrated or diluted when tomatoes are processed to juice, paste, ketchup, and other products needs to be known. It is clearly much easier and more reliable to have the analytical chemist measure the concentrations of residual fungicide in the tomatoes and tomato products than to estimate residue levels by calculation.

Such analytical measurements are necessary to establish concentrations for most agents in the environment. How much benzene is present in the air at gasoline stations as a result of its evaporation from gasoline? What is the concentration of arsenic in water running off the surface of a hazardous waste site where unknown amounts of arsenic were buried over many years? What is the polychlorinated biphenyl (PCB) concentration in fish swimming in waters next to a hazardous waste site known to contain this substance? How much aflatoxin is in a batch of peanut butter? The most reliable answers to these questions are those resulting from chemical analysis.

Analytical chemistry has undergone extraordinary advances over the past three to four decades. Chemists are able to measure many chemicals at the parts-per-billion level which in the 1960s could be measured only at the parts-per-million level (that's 1000 times more concentrated), or even the parts-per-thousand level. Chemical measurements now often reach to the parts-per-trillion level. (Even more dilute concentrations can be measured for some substances, but generally not for those in the environment.) These advances in detection capabilities have revealed that industrial chemicals are more widespread in the environment than might have been guessed 30 years ago, simply because chemists are now capable of measuring concentrations that could not be detected with analytical technology available in the 1970s. This trend will no doubt continue, because analytical chemists do not like to be told to stop developing the technology needed to search for smaller and smaller concentrations of chemicals: "chasing zero" is a challenge chemists are trained to pursue, even though they all know "zero" can never be reached.

While analytical science is in many ways quite miraculous, it is by no means without problems. Errors can easily be made. Analyses are not always readily reproducible in different laboratories. Some technologies are exceedingly expensive. And while analytical methods are well worked out for many chemicals, they are not available at all for many more. (Indeed, if we are interested in the naturally occurring chemicals that human beings are exposed to, we will find that only a tiny fraction of these can now be analyzed for with anything except fairly sophisticated research tools; most such chemicals are still unknown, and it is virtually impossible to develop routine and reliable analytical methods for chemicals that have not even been characterized.)

A third way to gain some knowledge about the concentrations of chemicals in the environment involves some type of modeling. Scientists have had, for example, fair success in estimating the concentrations of chemicals in the air in the vicinity of facilities that emit those chemicals. Information on the amount of chemical emitted per unit time can be inserted into various mathematical models that have been designed to represent the physical phenomena governing dispersion of the chemical from its source. Certain properties of the chemical and of the atmosphere it enters, together with data on local weather conditions, are combined in these models to yield desired estimates of chemical concentrations at various distances from the source. These models can be "calibrated" with actual measurement data for a few chemicals, and then used for others where measurement data are not available.

## Sampling the environment

Another uncertainty in understanding environmental concentrations arises because of the problem referred to as *sampling*. Suppose inspectors from the FDA want to know whether a shipment of thousands of heads of lettuce from Mexico contains illegal concentrations of a particular pesticide. Obviously, the entire shipment cannot be sampled, because analysis destroys the lettuce. So the inspector takes a few heads from different areas of the shipment and these are either combined in the laboratory (a "composite" analysis) or analyzed individually. In either case a certain concentration (for the composite) or a range of concentrations (for the individual heads) is reported from the laboratory. How can the inspector be sure these results fairly represent the entire shipment?

Suppose EPA scientists are investigating lead contamination of soil near an abandoned mining and smelting operation. How should the soil be sampled so that the analytical results obtained on the individual samples provide a reliable indication of the range and distribution of contamination of the entire site?

In neither the case of the pesticide residues in lettuce nor the lead concentrations in soil can it be certain that the whole is accurately represented by the part taken for analysis. Scientists can only have various degrees of confidence that the samples taken are representative. Statisticians can devise sampling plans that, if followed, allow scientists to know the degree of confidence, but that's the best that can be done. In practice much environmental sampling is done without a well-thought-out plan, in which case no statement can be made about the degree to which the whole is represented by the part. This unfortunate and unscientific practice complicates the lives of decision-makers, who many times have no recourse but to ignore the problem. (There are occasions when "statistical representativeness" is not very important. If we suspect the presence of some serious and potentially life threatening contamination we want to learn about it as quickly as possible; here we sample not to understand everything, but to learn whether some emergency action is needed.)

## Overview of dose estimation

The scientific issues associated with understanding and estimating human exposure to chemicals in the environment are vastly more complicated than has been suggested in this chapter, but for present purposes they have been covered sufficiently. The primary purpose of the discussion is to introduce some terms that will come up frequently in later chapters, and to provide some insight into how scientists come to understand how much of which chemicals are present in environmental media, and some of the ways they can come to enter the human body.

The dose resulting from exposure can be quantitatively estimated if we have knowledge of the amount of chemical present in a given weight or volume of medium, the weight of medium ingested or the volume inhaled per unit of time, and the body weights of the individuals who are exposed. In the typical regulatory risk assessment, the analysis is focused on the "high end" of the range of possible doses, to ensure the risk assessment results reflect those at highest risk.

We note here, and will discuss more fully below, the usefulness of having more refined measures of dose than the one discussed so far.

We are coming close to the central topic of toxicity, which obviously cannot occur until chemicals actually contact various parts of the body. But there is one more exceedingly important step that needs to be examined before toxic effects are considered. How do chemicals enter, move around within, and exit the body? With this subject we begin to understand the relationship between the doses the body experiences from contact with environmental media to those doses that come to be present in different fluids, organs, and cells of the body.

## Into the body

Chemicals in the environment may enter the mouth and be swallowed into the gastrointestinal tract. If they are in the vapor state or are attached to very fine dusts in the air, they may be inhaled through the nose and mouth and thereby enter the airways leading to the lungs. Some chemicals reach the skin, sometimes dissolved in some medium, sometimes not. What happens following the contact of environmental chemicals with these three routes of entry to the body?

First, chemicals come into intimate contact with the fluids, tissues, and cells that make up these three passages into the body. This contact may or may not result in some type of injury to tissues and cells; if some adverse response occurs in the tissues comprising these entryways, it is referred to as *local* toxicity.

In most cases, however, chemicals enter the bloodstream after they are *absorbed* through the walls of the gastrointestinal tract or the lungs, or through the layers that make up the skin. Once in the bloodstream they can be *distributed* through the body and reach the tissues and cells that make up the many organs and systems of the body. In most cases chemicals undergo molecular changes – chemical reactions in the cells of the body's organs, particularly the liver but in others as well: they are *metabolized*. Metabolism is brought about by enzymes – large protein molecules – that are present in cells. Chemicals and their metabolites (the products of metabolism) are then *eliminated* from the body, typically in urine, often in feces, and in exhaled air, and sometimes through the sweat and saliva.

The nature of toxic damage produced by a chemical, the part of the body where that damage occurs, the severity of the damage, and the likelihood that the damage can be reversed, all depend upon the processes of absorption, distribution, metabolism and elimination, ADME for short.

The study of ADME had its origins in pharmacology, the science of drugs. Because these processes involve *rates* of different types, this area of study came to be called *pharmacokinetics*.[2] The combined effects of these pharmacokinetic processes determine the concentration a particular chemical (the chemical entering the body or one or more of its metabolites) will achieve in various tissues and cells of the

---

[2] The same types of studies involving substances that are not pharmaceuticals, and that may produce toxicity, have been labeled by some as toxicokinetic studies. Because of a personal dislike of the latter term I shall continue to use the term pharmacokinetics in this book.

body and the duration of time it spends there. Chemical form, concentration, and duration are critical determinants of the nature and
extent of toxic injury produced. Injury produced after absorption is
referred to as *systemic toxicity*, to contrast it with local toxicity.

Here we see new measures of dose, and measures that are more
likely to have a direct bearing on toxic responses than what is often
called the administered dose (the doses of aspirin or TCE calculated
earlier in this chapter).

First, a word about how pharmacokinetic studies are typically performed. A highly interesting technique is used to follow a chemical
through the body. Chemists have learned to increase the natural level
of certain radioactive forms, or isotopes, of carbon, hydrogen, and
some other atoms in organic compounds. They can, for example, create molecules enriched in carbon-14, a radioactive isotope having two
more neutrons in the atom's nucleus than does the most abundant,
non-radioactive isotope of carbon. The ADME pattern of the radioac-
tively labelled chemical can easily be traced – indeed, these are called
radioactive tracer studies – because it is relatively simple to locate in the
body the extra radioactivity associated with the added carbon-14. The
presence of extra amounts of the radioactive isotope of carbon does
not alter in any significant way the chemical behavior of the molecule
carrying it. Radiotracer studies have not only made possible modern
toxicology, but almost all of modern biochemistry and pharmacology
depend upon their use. The Hungarian scientist who developed the
technique, Georg de Hevesy, was awarded the Nobel Prize in 1943,
although its full importance was realized only in later decades.

Those substances in the environment that are essential to the health
of the organism – the nutrients, oxygen, water – all undergo ADME.
Everyone is familiar with the fact that carbohydrates, proteins, and
fats break down in the body – the initial stages occurring in the gastrointestinal tract, and the rest in the liver and other organs after
absorption – and that the chemical changes these essential chemicals
undergo (their metabolism) are linked to the production and storage
of energy and the synthesis of the molecules that are essential to life.
Everyone is also familiar with the fact that there are variations among
people in the ease with which they can absorb certain nutrients – iron
and calcium, for example – from the gastrointestinal tract into the
blood. Although the chemicals we shall be discussing are not essential
to life, and can be quite harmful under some conditions, our bodies have mechanisms for absorbing, distributing, metabolizing, and
excreting them. These mechanisms probably evolved because humans

and other mammals have always been exposed to huge numbers of non-essential natural chemicals that have always been constituents of their foods and, to a lesser extent, of the air they breathe and water they drink.

# Absorption

### Gastrointestinal tract

Figure 2.1 depicts in schematic form the relationships among certain organs and systems of the body essential for an understanding of ADME. The arrows in the figure depict the various paths chemicals follow when they enter the body, move around within it, and are finally excreted from it. It will be helpful to refer to the figure throughout the course of this chapter.

It is not an oversimplification to describe the gastrointestinal (GI) tract as one very long tube open at both ends, the mouth and the anus. The major pieces of the tube are the mouth, throat, esophagus, stomach, small intestine and large intestine, or colon, and the rectum and anus. Most of the length is due to the intestines, which are actually highly coiled.

Chemicals in food and water, some medicines, and even some present in soils or dusts that are incidentally ingested can be absorbed along the entire GI tract. By absorption we mean the movement of the chemical through the membranes of the different types of cells comprising the wall of the GI tract so that it ends up in the bloodstream.

There are several different biological mechanisms at work to effect absorption, but their nature need not concern us here. Suffice it to say that a host of factors affect the site along the GI tract where a chemical is absorbed and the rate and extent of its absorption. Among these are the particular chemical and physical properties of the chemical itself, the characteristics of the medium – food (even the type of food) or water – in which it enters the GI tract, and several factors related to the physiological characteristics of the exposed individual. Toxicologists refer to the latter as *host* factors, because they belong to the individual that is playing "host" to the entering chemical.

Certain drugs are administered as sublingual tablets (they are placed under the tongue) and as rectal suppositories; these substances can be absorbed in the mouth and rectum, respectively.

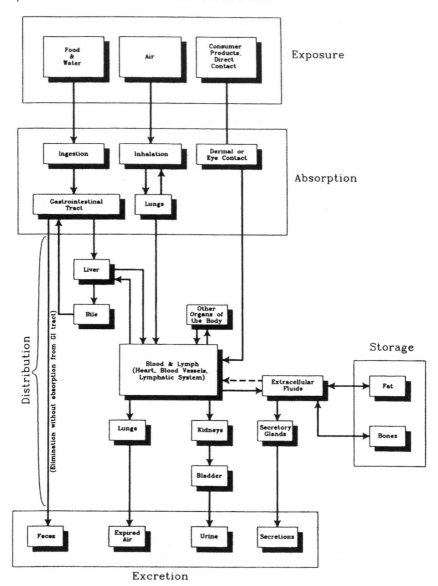

*Figure 2.1  A schematic showing how chemicals may enter, be absorbed into, distributed within, and excreted from the body.*

These two GI tract sites appear to be minor routes of entry for most environmental chemicals; the latter are more typically absorbed through the walls of the stomach and intestines.

Chemicals vary greatly in the extent to which they are absorbed through the walls of the GI tract. At one extreme are some very inert and highly insoluble substances – sand (silicon dioxide) and certain insoluble minerals such as several of the silicates added to foods to keep them dry – that are almost entirely unabsorbed. Such substances simply wind their way down the entire length of the GI tract and end up excreted in feces. This pathway is shown in Figure 2.1 as the long arrow extending from the GI tract directly to feces.

Most substances are absorbed to a degree, but few are entirely absorbed. Lead absorption from food, for example, may be in the range of 50%, but is less when this heavy metal is in certain highly insoluble chemical forms, or when it is associated with certain media such as dried paint or soils.

Much else is known about GI absorption. Individuals vary in the extent to which they can absorb the same chemical, and absorption can be influenced by individual factors such as age, sex, health status, and even dietary habits. People who consume large amounts of fiber may absorb less calcium and iron than those who eat less. The GI tract is not fully developed until about 24 months after birth, and infants absorb metals such as lead and certain organic chemicals more readily than do adults.

Different animal species exhibit differences in GI absorption rates. The extent of GI absorption of lead in rats, for example, can be studied by feeding the animals known amounts of the metal and analyzing the unabsorbed amount that comes through in feces; the difference is the amount absorbed. But because of possible species differences it is not possible to conclude that humans will absorb the same amount of lead as the rat. These types of differences complicate evaluation of toxic potential. At the same time, they help to explain why different species of animals respond differently to the same dose of a chemical.

### Respiratory tract

The respiratory tract includes the air passages through the nose and mouth that connect to the tubing (bronchi) that lead to the lungs. Gases such as oxygen and carbon dioxide, and the environmentally important pollutants carbon monoxide, nitrogen oxides, sulfur dioxide, and ozone, can readily travel the length of the respiratory tract and enter the lungs. So can vapors of volatile liquids such as gasoline and certain solvents. Sometimes such chemicals cause local toxicity – everything from minor, reversible irritation of the airways to serious, irreversible injury such as lung cancer – but, as with the GI tract, some amounts of these agents pass through the lungs (in the so-called alveolar area) and into the blood. The rate and extent of lung absorption are influenced by a number of host factors, including physiological variabilities specific to the animal species and even to individuals within a species.

Dusts in the air can also enter the airways. Here the physical dimensions of the individual particles determine the degree to which they migrate down the respiratory tract and reach the lung. Generally, only very fine particles, those smaller than about one micron (one-millionth of a meter), enter the deep (alveolar) region of the lung. Larger particles either do not enter the respiratory tract or are trapped in the nose and excreted by blowing or sneezing. Some particles deposited in the upper regions of the respiratory tract may be carried to the pharynx and be coughed up or swallowed. Thus, inhaled chemicals or dusts can enter the body by the GI tract as well as the respiratory tract.

Dusts may carry chemicals into the lungs, where they can be absorbed by several mechanisms. But there are other physical materials – asbestos is the most well known – that, depending upon their physical dimensions, can also be inhaled and can move down the respiratory tract to the lung, where they can reside in insoluble form for long periods of time and cause serious, lasting damage.

## Skin

The skin acts as a barrier to the entry of chemicals, but some chemicals get through it. Dermal, or percutaneous absorption, as it is technically called, generally involves diffusion of a chemical through the so-called epidermis, which includes the outer layer of dead cells called the stratum corneum. This is a tough barrier for chemicals to get through, and many don't make it. If they do, they also have to negotiate passage through the less protective second layer called the dermis; once past this they are in to the blood.

The effectiveness of the stratum corneum in blocking the passage of chemicals varies from one part of the human body to another. It is particularly difficult for chemicals to cross the palms of the hands and soles of the feet, but they get by the scrotum fairly easily. Abdominal skin is of intermediate effectiveness in preventing absorption. Age and sex also influence rates of dermal absorption, and some species of animals (e.g. rabbits) seem to have much more vulnerable skins than others; humans and other primates appear to be near the least vulnerable end of the scale.

Not surprisingly damage to the skin enhances absorption rates; if the less protective dermis is exposed because the stratum corneum has been scraped off, penetration can be substantial.

The physical properties of a chemical, which are in turn functions of its chemical structure, have a powerful influence on its likelihood of getting through the skin. Generally, chemicals must be capable of dissolving fairly readily in both water and fat-like materials. Substances that dissolve only in water and those that have little affinity for water but only for fatty materials, do not get far. Large molecules cannot move as easily through the skin as can smaller ones. Substances that do not dissolve well in water or any other solvent just cannot penetrate in measurable amounts.

## Distribution

Once absorption occurs and a chemical is in the blood it can move around the body with relative ease (see Figure 2.1), going almost everywhere blood goes. It generally will not be distributed equally to all organs and systems; and the pattern of distribution will vary greatly among chemicals, according to their particular structural characteristics and physical properties. Most chemicals undergo reversible binding to proteins present in blood; often the binding reduces the amount of chemical available for entry into tissues.

There are fortunately a few other natural biological barriers that prevent or impede distribution of chemicals to certain organs. The most important of these are the *blood–brain* barrier and the *placental* barrier, the one retarding entry of chemicals to the brain, the other protecting the developing fetus. These barriers are not perfect, however, and certain chemicals can migrate through them. Most chemical forms of the metal mercury, for example, cannot readily pass the blood–brain barrier, and they exert their primary toxic effects on the kidney, not the brain. But there is a certain chemical form of mercury, called methylmercury, that can break through the barrier, mostly because it can dissolve in fatty materials while the other forms of mercury cannot, and this form can cause damage to the brain. Methylmercury can also invade the placental barrier while other forms of mercury are largely locked out. Other biological barriers, none perfect, exist in the eye and testicles.

Certain chemicals can be *stored* in the body, as depicted in Figure 2.1. A major site of storage is bone, which can bind metals such as lead and strontium and non-metallic inorganic elements such as fluoride. While bound in this form the chemicals are relatively inert, but under certain conditions, they can be slowly released from storage and re-enter the bloodstream where they are more available to cause biological effects.

Another tissue – fat – can store certain organic chemicals that are highly soluble in this medium. Certain pesticides such as DDT and industrial products such as polychlorinated biphenyls (PCBs) readily dissolve in body fat and can stay there for long periods of time. Most people have measurable amounts of these two once widely used chemicals, and several more as well, in their fat stores.

Chemicals stored in fat and bone are at the same time also present in blood, usually at very much lower concentrations. A kind of equilibrium exists between the storage tissue and blood. If more of the chemical is absorbed into the body, its blood concentration will first rise, but then will fall as some of it enters the storage area; equilibrium conditions are eventually re-established, with higher concentrations of the chemical in both the medium of storage and the blood. Likewise, removal of the source of exposure and loss of the chemical from blood through excretion (see below) will mobilize the chemical from its storage depot and send it into the blood. As long as no external sources of the chemical are available, it will continue to be lost from the blood and continue to migrate out of storage. Finally, it will all but disappear from the body, but this may take a very long time for some chemicals.

An interesting phenomenon has been observed in people who have lost weight. Removal of body fat decreases the amount available for storage of fat-soluble chemicals. Blood concentrations of DDT have been observed temporarily to increase following weight loss. In light of the discussion above, such an observation is not unexpected.

## Metabolism

The cells of the body, particularly those of the liver, and with important contributions from those of the skin, lungs, intestines and kidneys, have the capacity to bring about chemical changes in a large number of the natural and synthetic chemicals that are not essential to life. As we have said, these chemical changes yield metabolites of the absorbed chemical, and the process whereby metabolites are produced is called metabolism. The latter term is also applied to the biochemical changes associated with nutrients and substances essential to life.

Often metabolites are more readily excreted from the body than the chemical that entered it, and chemical pathways leading to such metabolites are called detoxification pathways; the quicker the chemical is eliminated, the less chance it has to cause injury.

Metabolism is generally brought about, or catalyzed, by certain proteins called enzymes. Cells have many enzymes and most are involved in biochemical changes associated with their ordinary life processes. But some act on non-essential chemicals, on "foreign" compounds, and convert them to forms having reduced toxicity and enhanced capacity for excretion. Why we should have such enzymes is not clear, but some are probably the result of evolutionary adaptations that increased species survival in the face of environmental threats. These enzyme systems evolved over very long periods of time, primarily in response to millions of years of exposure of cells to naturally occurring chemicals, and they were thus available for the exposures to synthetic chemicals that began only about 125 years ago. Most synthetic chemicals have the same groupings of atoms that are found in naturally occurring molecules, although important differences certainly exist. In particular, the carbon–chlorine bond, common in some important industrial solvents and other chemicals, is relatively rare in nature. It is perhaps not surprising that so many chemicals that have prompted public health concerns contain carbon–chlorine bonds.

An example of beneficial metabolism is illustrated by the conversion of toluene, a volatile chemical present in petroleum products and readily absorbed through the lungs, to benzoic acid, as shown:

Toluene is chemically related to benzene, but a methyl ($CH_3-$) grouping has replaced one of the hydrogen atoms attached to the ring of six carbon atoms. The "circle" drawn inside the six-carbon ring represents a set of six electrons that comprises the chemical bonds involved in the so-called aromatic ring. This collection of six carbon atoms in a ring with six electrons represented by the circle appears in many organic compounds, including a number appearing later in this book.

At high exposures toluene molecules can reach and impair the nervous system. But the liver has certain enzymes that can eliminate the three hydrogen atoms attached to the carbon atom, those of the methyl group, and introduce oxygen atoms in their place. The rest of the toluene molecule is unaffected. But this metabolic change is enough. Benzoic acid is much less toxic than toluene – indeed, it has very low toxicity – and it is much more readily excreted from the body. This metabolic pathway detoxifies toluene. Of course, if an individual inhales huge amounts, such that the amount of toluene absorbed and distributed exceeds the capacity of the liver to convert it to benzoic acid, then toxicity to the nervous system can be caused by the excess toluene. Though toluene is chemically related to benzene, the latter is a far more toxic chemical – it can cause certain blood disorders and forms of leukemia in humans – in part because it cannot be so readily metabolized to chemical forms having reduced toxicity.

Toxicity can occur because, unfortunately, some metabolites are, unlike benzoic acid, more toxic than the chemical that enters the body. Enzymes can cause certain changes in molecular arrangements that introduce groupings of atoms that can interact with components of cells in highly damaging ways. The industrial chemical bromobenzene can be converted in the liver to a metabolite called bromobenzene epoxide, as depicted in the diagram.

Bromobenzene

liver cells

Bromobenzene epoxide
(highly reactive)

Path A

Reacts with liver cell molecules leading to cell death

Path B
(detoxification)

Creation of several molecules that are readily excreted from the body

The epoxide molecule is very active and can bind chemically to certain liver cell molecules and cause damage and even death to the cell (Path A). But an alternative reaction path (Path B) can also operate. If the amount of bromobenzene that enters the cell is low enough, Path B (which actually creates several metabolites) dominates and little or no cell damage occurs because the metabolic products are relatively non-toxic and are readily excreted from the body. But as soon as the capacity of the cell to detoxify is overcome because of excessive concentrations of bromobenzene, the dangerous Path A begins to operate and cell damage ensues.

Metabolism of organic molecules often occurs in two phases. Phase I generally involves the conversion of certain functional (chemically reactive) groups in the molecule from non-polar (lipid-soluble) to polar (having affinity for water) groups. Phase I reactions, which may increase, decrease, or leave unaltered a compound's toxicity, are catalyzed by a system of enzymes called cytochrome P-450 (CYP). The CYP system is important for the metabolism of both endogenous compounds (those natural to the body, such as steroidal hormones or lipids) and for foreign chemicals. The CYP enzymes are present in most cells, but are most active in the liver and intestinal tract. There are a number of subfamilies of these enzymes; specific enzymes within those subfamilies are called isozymes (designated with the addition of letters and numbers – CYP3A4, CYP2D6, etc.). Interestingly and importantly, some chemicals or drugs can *induce the synthesis* of one or more isozymes, and thereby affect the metabolism (and subsequent elimination rates) of other chemicals or drugs. Such effects on metabolism and clearance rates can lead to increases in the levels of toxic metabolites or delay their elimination, and thereby increase toxicity. Of course the opposite effects can also occur, depending upon which isozymes are induced.

The metabolites resulting from Phase I reactions may be sufficiently water soluble to undergo renal elimination. But if not, there is a second metabolic phase available. Phase II reactions are referred to as *conjugation reactions*, because the polar group (typically $-OH$, $-NH_2$, or $-COOH$) created in Phase I undergoes bonding (conjugation) with certain acidic endogenous compounds (glucuronic acid, some amino acids, acetic acid) to create highly water-soluble, and readily excretable, conjugates. Perhaps glucuronides are the most important of the conjugates created in Phase II reactions.

Toxic metabolites are common and toxicologists are learning that many if not most types of toxic, and even carcinogenic, damage are actually brought about by metabolites. Additional examples of this phenomenon surface in later chapters.

As with absorption and distribution, the nature and rate of metabolic transformations vary among individuals and different animal species. Metabolism differences can be extreme, and may be the most important factor accounting for differences in response to chemical toxicity among animal species and individuals within a species. The more understanding toxicologists acquire of metabolism, the more they shall understand the range of responses exhibited by different species and individuals, and the better they shall be able to evaluate toxic risks to humans.

## Elimination

Most chemicals and their metabolites eventually depart the body; they are eliminated (or excreted), as shown in Figure 2.1. The speed at which they leave varies greatly among chemicals, from a few minutes to many years. Because a chemical leaves the body quickly does not mean it is not toxic; damage to components of cells from certain chemicals can occur very quickly. Likewise, because a chemical is poorly excreted does not mean it is highly toxic, although it is true that long residence times can increase the chance of an adverse event occurring.

Elimination rates are commonly expressed as half-lives, the time required for half the amount of a chemical to leave the body. Half-lives for rapidly eliminated chemicals are typically in the range of a few hours. Highly persistent chemicals (see below) have half-lives of years.

As blood moves through the kidneys, chemicals and their metabolites can be filtered out or otherwise lost from the blood by a set of extraordinary physiological mechanisms that release them into urine. Urinary excretion is probably the pathway out of the body for most chemicals.

Gases and highly volatile chemicals can move out of the blood into the lungs, and be exhaled. Carbon dioxide, for example, is a metabolic product of many chemicals and also derives from the metabolism of essential molecules; it is excreted from the body through the lungs.

Chemicals may also be excreted in bile. Bile is a fluid normally excreted by the liver. It is composed of some degradation products of normal metabolism, and is excreted out of the liver and into the GI tract. Some chemicals move into bile, out into the GI tract, and are then excreted in feces (along with chemicals that are not absorbed from the GI tract, as discussed in the Absorption section earlier).

Interestingly, some metabolites undergoing biliary excretion are reabsorbed, usually after undergoing further metabolic change brought about by enzymes associated with microorganisms normally found in the intestines. There are notable examples of this phenomenon, and it can be important as a factor in toxicity production, but its discussion is beyond the scope of this book.

Some minor routes of excretion exist: sweat, hair, saliva, semen, milk. While these routes out of the body do not count for much as excretory processes, excretion of some chemicals into milk can be important because it constitutes an *exposure pathway* for infants, if the milk is from their mothers, and for many people if it is from dairy cattle. Many fat-soluble chemicals follow this pathway out of the body, dissolved in the fatty portion of the milk. Excretion of chemicals through milk is common enough to prompt considerable attention from toxicologists.

## Direct measures of dose

In recent years there has been an upsurge in efforts to move from indirect measures of human exposure (obtained by measurements of chemicals in environmental media and estimation of dose accrued from contact with those media) to direct measures of the concentrations of chemicals in the body, typically in blood and in elimination pathways such as urine and hair. The most significant effort in this direction in the United States has been undertaken as part of the CDC's National Health and Nutrition Examination Survey (NHANES).

This survey is, as its name suggests, a national, statistically based survey devoted to the development of information on the health and nutritional status of the population. The first survey, heavily oriented to questions of nutrition, was undertaken from 1971–1975. In the most recent national survey (1999–2001) blood and urine samples were collected in close to 8000 children and adults and were then analyzed for 116 chemicals. The survey focused heavily on commonly used pesticides and a number of important environmental contaminants, particularly those that persist in the body for long periods of time. Not surprisingly, most of the 116 were detected in at least some individuals. Such findings are not in the least surprising, because it is obvious that all of us are exposed to substantial numbers of chemicals and that some persist in the body for long periods of time. More NHANES surveys are underway, and similar efforts are underway in research organizations all over the globe.

That some chemicals persist for long periods has been recognized for many years. Most of the pesticidal chemicals that came into wide use in the 1930s and 1940s were chlorinated organic compounds such as DDT, dieldrin, chlordane, kepone, toxaphene, and several others.

1,1,1-Trichloro-2,2-bis(p-chlorophenyl)ethane (DDT)

Polychlorinated biphenyl (PCB)

3,4,4',5'-Tetrachlorobiphenyl

Polychlorinated dibenzo-p-dioxin (PCDD)

2,3,7,8-Tetrachlorodibenzo-p-dioxin (2,3,7,8-TCDD)

In the United States most uses of these chemicals ended in the 1970s, and knowledge that they degraded in the environment at very slow rates was, along with their adverse effects on wildlife, the principal reason government agencies around the world have acted to restrict or, in most cases, to ban their use. Newer generations of pesticides and those now in use are generally not persistent (though the use of all such "economic poisons" remains controversial). These chlorinated organic pesticides and other chlorinated organic chemicals that once had wide industrial use (polychlorinated biphenyls, PCBs) or that are created as by-products of certain industrial processes or incineration (polychlorinated dioxins) were detected in the NHANES survey, and have been reported in many other scientific studies of humans and wildlife. In recent years industrial products and by-products that contain many carbon–fluorine and carbon–bromine bonds have also been detected in humans and other animal species. The bonds between carbon and fluorine, chlorine, and bromine are, when the carbon is part of an aromatic ring (as in PCBs and dioxins), very strong and unreactive (they are not readily ruptured by either chemical or enzymatic processes); even some polyfluorinated organic compounds that are not aromatic are highly stable. High chemical stability and a high affinity for fat contribute to the persistence of these compounds, not only in the human body, but everywhere in the living environment.

The so-called persistent organic pollutants (POPs) are the subject of the Rio Declaration (1994) and the Stockholm Convention (2001); these international agreements (yet to be ratified in all signatory countries, including the United States) call for the elimination from production of 12 persistent chemicals, including the chlorinated pesticides and PCBs mentioned above (all of which have already been eliminated from production in the United States). Current regulatory efforts in the European Union and the United States place emphasis on elimination or restriction of all PBTs (persistent, bioaccumulative, and toxic chemicals).

Knowledge of the levels of chemicals in the body can be used to track the effectiveness of efforts to reduce human exposure to certain chemicals. It may be of use in the conduct of health studies and risk assessments. The NHANES survey and similar efforts around the world will, in the next several years, greatly increase our understanding of chemical exposure.

Because there are limited scientific tools now available to understand the health risks associated with various levels of chemicals in the body, it is likely that these results will also increase public concern. We should emphasize that persistence in the body is by no means a certain predictor of toxicity. Some chemicals may persist for long periods without causing harm, because the concentrations are low and because they have little of the type of chemical reactivity needed to initiate toxic events. And some chemicals that are eliminated from the body very quickly may leave behind significant injury. This general knowledge is, however, probably inadequate to allay developing public concern about highly persistent substances.

## Uses of pharmacokinetic data

Toxicologists generally believe that comprehensive pharmacokinetic data can provide extraordinarily useful information to assist in judging the risks posed by chemical agents. The reasons for this belief are complex, and the best we shall be able to do in this book is to illustrate some of the uses of these data when we discuss specific toxic agents. Up until now our purpose has been simply to create some understanding of how chemicals get into, move about, and get eliminated from the body, how they undergo chemical change, and to suggest why these processes are important determinants of toxicity.

A very important issue has not been mentioned throughout this discussion: it is generally not possible to acquire comprehensive pharmacokinetic data in human beings! To study pharmacokinetics systematically, and to develop reliable data, requires studies of a type that simply cannot be performed ethically in human beings, at least with chemicals of more than a very low toxic potential or having clear health benefits, as might be the case with certain drugs. It is possible to acquire a little ADME data by, for example, conducting careful chemical analysis on exhaled air or urine of people (sampling of other fluids or tissues, except perhaps for blood, obviously cannot be routinely performed) known to be exposed to certain chemicals, say in the workplace. Such analyses may reveal the amounts and chemical identities of excreted metabolites, but little else. Such information may nevertheless be useful, because toxicologists can compare the pattern of urinary metabolites observed in humans and in experimental animals. If the patterns are similar, this tends to support the proposition that toxic effects observed in the species of experimental animals may be relevant to humans, whereas a substantially different pattern may suggest the opposite. In general, human ADME data, except for some pharmaceutical agents extensively studied in clinical settings, are fairly limited for environmental chemicals. Nevertheless, we shall see in later sections of this book, and especially in Chapter 9, some of the ways in which information on pharmacokinetics, when it can be developed, has the potential to improve substantially the scientific bases for risk assessment.

# 3

# From dose to toxic response

It is difficult to doubt that the earliest human beings, and perhaps even some of their evolutionary predecessors, were aware of poisons in their environments. For as long as human beings have walked the earth they have been stung or bitten by poisonous insects and animals. In their search for nourishing foods, mostly through trial and error, some early members of the species no doubt were sickened by or even succumbed to the consumption of the many plants that contain highly toxic constituents. Somewhere in prerecorded time, human beings also learned that certain plants could alleviate pain or remedy certain afflictions; learning about these plants probably also taught them a great deal about unpleasant side effects.

Those early metallurgists who were clever enough to learn how to transform crude ores to shiny metals were probably also observant enough to discover that some of the materials being worked with could harm them. Some of the earliest written accounts of humans on earth provide evidence that the ancient Greeks and Romans were well aware of the poisonous properties of certain plants and metals. The case of the poisoning of Socrates with hemlock is only the most famous of the early references to the deliberate use of certain plants for suicidal or homicidal purposes.

The science of toxicology, which we define as the study of the adverse effects of chemicals on health and of the conditions under which those effects occur, has begun to take on a well-defined shape only in the past four to five decades. The science is still struggling for a clear identity, but it has begun to find one. One of the reasons for its uncertain status is that toxicology has had to borrow principles and methodologies from several more basic sciences and from certain medical specialties. Several disparate historical strands of the study of poisons have intertwined to create the science of modern toxicology.

One strand, and probably the earliest to appear in a systematic form, was the study of antidotes. The modern strand of this field of study is called clinical toxicology. Clinical toxicologists are typically physicians who treat individuals who have suffered deliberate or accidental poisoning. Poisoning as a political act was extremely common up to and during the Renaissance, and we have witnessed a recent example in the dioxin poisoning of Ukrainian opposition candidate and now President Viktor Yuschenko.[1] Poisoning for homicidal purposes continues on a not insignificant scale to this day.

Some of the earliest physicians, including Dioscorides, a Greek who served Nero, and the great Galen himself, were engaged to identify ways to reverse poisonings or to limit the damage they might cause. The Jewish philosopher Maimonides published *Poisons and Their Antidotes* in Arabic, in the year 1312; this text synthesized all knowledge available at the time and served as a guide to physicians for several centuries. The Spanish physician Mattieu Joseph Bonaventura Orfila published in 1814–15 a comprehensive work entitled *A General System of Toxicology or, a Treatise on Poisons, Found in the Mineral, Vegetable, and Animal Kingdoms, Considered in Their Relations with Physiology, Pathology, and Medical Jurisprudence*. Orfila's organization of the topic is considered a seminal event in the history of toxicology.

A second major strand, closely linked to the first, is in the domain of what is today called pharmacology: the study of drugs and of their beneficial and adverse effects. This strand is also a very ancient one. Pharmacologists, many of whom were also botanists, at first collected plants and made catalogues of their beneficial and harmful effects. Some of these works are magnificent compilations of highly detailed information and have proved to be of enormous benefit to humankind. Major advances in pharmacology were brought about by the work of Paracelsus (*c.* 1493–1541), a Swiss physician and alchemist who promoted theories of disease that were an odd mix of scientifically

---

[1] At high doses dioxin causes a serious form of acne (chloracne), which appears to be Yuschenko's problem. If the poisoner's intention was to kill him, then he failed to select the right chemical. There are many far more effective acute poisons, as we shall see in the next chapter.

advanced notions and fanciful superstitions. Among toxicologists and pharmacologists he is noted for his recognition that

All substances are poisons; there is none which is not a poison. The right dose differentiates a poison and a remedy.

This remark adorns the frontispiece of almost every toxicology text. It was only in the late nineteenth and early twentieth centuries, however, that pharmacologists began to acquire some understanding of the nature of the specific chemical constituents of plants that were biologically active, in both beneficial and harmful ways. Once chemical science had undergone its revolution, it was possible for pharmacologists to begin to understand how molecular structure influenced biological action.

The tools for the systematic study of the behavior of these drug molecules in biological systems also came under rapid development during this same period, as the science of experimental medicine began to blossom. Pharmacologists drew upon the advances in medicine, biochemistry, and physiology resulting from the work of the great French physiologist Claude Bernard (1813–78) and began to create our modern understanding of drug action and drug toxicity. Some of the principal experimental tools used by modern toxicologists were first developed by pharmacologists. The methods used to collect ADME data, for example, were brought to perfection by scientists studying drug behavior. Some pharmacologists even contend that toxicology is merely a branch of pharmacology.

A third historical strand that has helped to create modern toxicology consists of the labors of occupational physicians. Some of the earliest treatises on toxicology were written by physicians who had observed or collected information on the hazards of various jobs. The man some have called the father of the field of occupational medicine was Bernardino Ramazzini, an Italian physician whose text *De Moribus Artificum Diatriba* (1700) contributed enormously to our understanding of how occupational exposure to metals such as lead and mercury could be harmful to workers. Ramazzini also recognized that it was important to consider the possibility that some poisons could slowly build up in the body and that their adverse effects do not make themselves apparent for a long time after exposure begins.

Sir Percival Pott published in 1775 the first record of occupationally related human cancers; this London physician recognized the link between cancer of the scrotum and the occupation of chimney sweep. More of Sir Percy's work will be described in Chapter 5.

Occupational physicians, Pott among them, contributed greatly to the development of the modern science of epidemiology – the systematic study of how diseases are distributed in human populations and of the factors that cause or contribute to them. Epidemiology is an important modern science, and its application, as we shall see, can provide the most significant data obtainable about the toxic effects of chemicals.

Occupational toxicology and industrial hygiene took a great leap forward in the early part of this century when Alice Hamilton (1869–1970), a physician from Fort Wayne, undertook with enormous energy and unyielding commitment an effort to call national attention to the plight of workers exposed to hazardous substances in mines, mills, and smelting plants throughout the country. Her work, which has perhaps been insufficiently acclaimed, led to renewed interest in occupational medicine and was also instrumental towards the introduction of worker's compensation laws. Dr. Hamilton became in 1919 the first woman to receive a faculty appointment at Harvard University.

Observations from the field of occupational medicine also created interest among biologists in the field of experimental carcinogenesis – the laboratory study of cancer development – that is the topic of Chapters 5 and 6.

Studies in the science of nutrition make up another strand leading to modern toxicology. The experimental study of nutrition, another offspring of the explosion in experimental medicine that took place following the work of Claude Bernard led, among other things, to increased appreciation of the proper use of experimental animals to understand the biological behavior of nutrients and other chemicals. The pioneering work on vitamins of Philip B. Hawk in the World War I era led to the development of experimental animal models for studying the beneficial and harmful effects of chemicals. One of Hawk's students, Bernard Oser, contributed enormously to the perfection of the experimental animal model, particularly in connection with the study of the toxicology of foods and food ingredients. Work in experimental nutrition also gave toxicologists a sense of the importance of individual and species variability in response to exogenous chemicals (i.e., chemicals entering the body from an outside source).

Modern toxicologists have also drawn upon the work of radiation biologists – scientists who study the biological effects of various forms of radiation. The development of radiation biology spawned important work on the genetic components of cells and the ways in which they might be damaged by environmental agents. It also provided insight into some of the biological processes involved in the development of cancers from damaged, or mutated, cells. The Manhattan Project itself created a need to understand the toxic properties of the myriad of chemicals that were then being prepared and handled in unprecedented amounts. Stafford Warren, head of the Department of Radiology at the University of Rochester, established a major research program on the toxicology of inhaled materials, including radioactive substances; Warren's group included Herbert Stokinger, Harold Hodge, and several other scientists who went on to become luminaries in the field.

For the past 100 years, and particularly since about 1935, toxicologists have been activated by developments in synthetic chemistry. As the chemical industry began to spew forth hundreds and thousands of new products, pressures were created for the development of information about their possible harmful effects. The first federal law that gave notice of a major social concern about poisonous products was the Pure Food and Drug Act, passed by Congress in 1906 and signed into law by Theodore Roosevelt. Much of the impetus for the law came from the work of Harvey Wiley, chief chemist of the Department of Agriculture, and his so-called "Poison Squad." Wiley and his team of chemists not infrequently dosed themselves with suspect chemicals to test for their deleterious effects. Wiley ran into trouble over the artificial sweetener, saccharin. The chemist thought it harmful; his boss Teddy Roosevelt, always somewhat overweight, was an advocate. Wiley somehow survived the spat, and his laboratory eventually evolved into the Food and Drug Administration.

The systematic study of toxic effects in laboratory animals began in the 1920s, in response to concerns about the unwanted side effects of food additives, drugs, and pesticides (DDT and related pesticides became available in this era).[2] Concerns uncovered during the 1930s and 1940s about occupational cancers and other chronic diseases resulting from chemical exposure prompted increased activity among toxicologists. The modern version of the Food, Drug, and Cosmetics Act was enacted by Congress in 1938 in response to a tragic episode in which more than 100 people died from acute kidney failure after ingesting certain samples of the antibiotic sulfanilamide ("Elixir of Sulfanilamide") that had been improperly prepared in a diethylene glycol solution. Diethylene glycol is obviously better suited for its use

---

[2] An interesting footnote to history: a Swiss physician, Paul Herman Müller, was awarded the Nobel Prize for medicine in 1948, for his work on DDT and its use in controlling insects that transmit malaria and typhus.

as antifreeze. This law was only the first of many that have contributed to the creation of the modern science of toxicology. The Elixir of Sulfanilamide tragedy had a beneficial side effect, in that it prompted some of the earliest investigations into underlying mechanisms of toxicity by Eugene Geiling at the University of Chicago. The scientists Geiling gathered to work on the diethylene glycol problem and on other emerging problems in toxicology were to become leaders of the field during the following three decades.

Sporadically during the 1940s and 1950s the public was presented with a series of seemingly unconnected announcements about poisonous pesticides in their foods (most infamous of which was the great cranberry scare of 1959, in which federal officials announced, just before Thanksgiving, that it would be "prudent" to avoid consuming these berries because they were contaminated with a carcinogenic herbicide), food additives of dubious safety, chemical disasters in the workplace, and air pollution episodes that claimed thousands of victims in urban centers throughout the world. In 1962 Rachel Carson, a biologist from Silver Spring, Maryland, drew together these various environmental horror stories in her book, *Silent Spring*. Carson wrote:

For the first time in the history of the world, every human being is now subjected to contact with dangerous chemicals, from the moment of conception until death. In the less than two decades of their use, the synthetic pesticides have been so thoroughly distributed throughout the animate and inanimate world that they occur virtually everywhere.

and

Human exposures to cancer-producing chemicals (including pesticides) are uncontrolled and they are multiple. . . . It is quite possible that no one of these exposures alone would be sufficient to precipitate malignancy – yet any single supposedly "safe dose" may be enough to tip the scales that are already loaded with other "safe doses."

*Silent Spring* was immensely popular and influential. Carson's work almost single-handedly created modern society's fears about synthetic chemicals in the environment and, among other things, fostered renewed interest in the science of toxicology. It also helped pave the way for the introduction of several major federal environmental laws in the late 1960s and early 1970s, and for the creation of the EPA in 1970.

Beginning in the 1930s individuals from a wide variety of scientific and medical disciplines, and working in various government, industry, and academic laboratories, began drawing upon the accumulated experience of the many fields of study that had been devoted to understanding the behavior of chemical substances, including some having radioactive properties, when they came into contact with living systems. Other than the physicians who had specialized in clinical toxicology and medical forensics, these individuals were not "toxicologists" – the discipline did not exist. They were pharmacologists and biochemists, physiologists, general biologists, epidemiologists, experimental nutritionists, pathologists, and scientists involved in experimental cancer studies. Scientists from the basic disciplines of chemistry and physics, and statisticians specializing in biological phenomena, also made contributions.

Out of this collective effort, a coherent scientific enterprise began to take shape. The first major professional group under which these investigators collected and published their work was the Society of Toxicology, founded as recently as 1961. A few graduate schools and schools of public health began offering advanced degrees in toxicology in the 1960s, and toxicology courses began to be included in other, related graduate curricula. Professional journals began to multiply and several are now internationally prominent. The National Academy of Sciences began to deal with toxicology issues in the 1940s, and they have since become a major feature of Academy efforts. These trends, and several others to be mentioned later in the book, have led to the common acceptance of a set of definitions, principles, and methodologies that guide the discipline. While considerable debate on certain matters exists, consensus or near-consensus exists on many others. The principal purpose of the remainder of this chapter is to set forth some of the important definitions, principles, and methodologies that have emerged and become firmly established over the past few decades. This discussion will set the stage for the later chapters, in which specific types of toxicity and toxic agents are examined.

## Activities of toxicologists

In brief, toxicologists are involved in three types of activity:

(A) The study of the types of adverse health effects produced by chemicals under various conditions of exposure. Epidemiologists are similarly engaged in their studies of exposed human populations. Toxicologists are usually in the laboratory, carrying out experiments in animals and other test systems.

(B) The study of the underlying biological events, including but not limited to ADME studies, by which chemicals create adverse health effects.
(C) The evaluation of information on the toxic properties of chemicals, the conditions under which these properties manifest themselves, and the underlying biological processes leading to toxicity, to assess the likelihood that those chemicals might produce their adverse effects in human populations that are or could be exposed to them.

Toxicologists engaged in activities (A) and (B) are typically laboratory scientists. Even today many involved in these efforts were not trained as toxicologists, but have come to them from a variety of disciplines. Moreover, some aspects of these activities require a degree of specialization, in disciplines such as pathology and statistics, that most toxicologists do not have.

Toxicologists engaged in activity (C) do so outside the laboratory. They may undertake such activities as members of various expert committees, as employees of regulatory agencies, or as scientists in corporations who are responsible for giving advice to management on matters of chemical risk. Activity (C) is called risk assessment; it is a difficult, controversial, and unsettled area to which several later chapters are devoted. Toxicologists engaged in risk assessment are typically aided by epidemiologists, statisticians, experts in human exposure analysis, and other toxicologists whose principal occupations are activities (A) and (B).

It should be clear, even from this somewhat oversimplified picture of what toxicologists do, that they do not do it all alone; toxicology is a discipline, but a thorough evaluation of the risks of chemical agents requires a multidisciplinary effort, and even today it is not possible to clearly define the boundaries between toxicology and the several disciplines it draws upon.

## Some important terms and principles

All chemicals, natural and synthetic, are toxic – that is, they produce adverse health effects – under some conditions of exposure. It is incorrect (but I'm afraid very common) to refer to some chemicals as toxic and others as non-toxic. If this book teaches any lesson, it is that this notion is not correct.

Chemicals do, however, differ greatly in their capacity to produce toxicity. The *conditions of exposure* under which toxic effects are produced – the size of the dose, the duration of dosing needed, and even the route of exposure – vary greatly among chemicals. Moreover, the nature and severity of the toxic effects produced are also highly varied, and are different not only for different chemicals, but also for a single chemical as the conditions of exposure to it change.

A commonly used scheme for categorizing toxicity is based on exposure duration. Toxicologists generally seek to understand the effects of *acute, chronic*, and *subchronic* exposures. They attempt to learn for each of these three exposure categories the types of adverse effects a chemical produces, the minimum dose at which these effects are observable, and something about how these adverse effects change as the dose is increased.

Acute exposure involves a single dose. Toxicologists frequently refer to the immediate adverse consequence of an acute exposure as an "acute effect." Such events are also referred to as poisonings. This usage is not incorrect, but, as will be seen in a moment, it can be misleading when similarly applied to chronic exposures.

In studying the acute toxicity of a chemical, our interest is in understanding the dose that will lead to some harmful response and also to the most harmful one of all, death. We shall see that some chemicals produce toxicity or death after a single exposure at extremely low dose, while others do so only at doses that are so high they are nearly impossible to get into the body. Most chemicals fall between these extremes. The notion that the world consists of two neatly separated categories of chemicals, the toxic and the non-toxic, derives largely from the notion that for many chemicals (the "non-toxic" ones) extremely high and unlikely doses are needed to produce acute toxicity. This toxic/non-toxic dichotomy, while as a practical matter useful for separating the substances we should be concerned about for their acute effects from those we need not worry about, can create a misleading impression about the nature of chemical risks.

Chronic exposure generally refers to repeated dosing over a whole lifetime, or something very close to it. Subchronic is less well-defined, but obviously refers to repeated exposures for some fraction of a lifetime. In animal toxicity studies involving rodents, chronic exposure generally refers to daily doses over about a two-year period, and subchronic generally refers to daily doses over 90 days. Again, for both these exposure durations, toxicologists are seeking to learn the specific types of adverse effects associated with specific doses of the chemical under study. Some dosing regimens do not fall usefully into the chronic or subchronic categories, and some of these will be encountered in later discussions of the effects of chemicals on the reproductive process or on fetal development.

Care must be taken to distinguish subchronic or chronic exposures from subchronic or chronic effects. By the latter, toxicologists generally refer to some adverse effect that does not appear immediately after exposure begins, but only after a delay; sometimes the effect may not be observed until near the end of a lifetime, even when exposure begins early in life (cancers, for example, are generally in this category of chronic effects). But the production of chronic effects may or may not require chronic exposure. For some chemicals acute or subchronic exposure may be all that is needed to produce a chronic toxicity; the effect is a delayed one. For others chronic exposure may be required to create chronic toxicity. Toxicologists are not always careful to distinguish between subchronic and chronic exposures and effects.[3] In this book we shall refer to exposures as subchronic or chronic, and talk about toxic effects as immediate (quickly following exposure) or delayed.

Toxicologists refer to *targets* of toxicity. Some chemicals damage the liver, others the kidney, and some damage both organs. Some adversely affect the nervous system, or the reproductive system, or the immune system, or the cardiovascular system. The brain, the lungs, elements of the blood, the blood vessels, the spleen, the stomach and intestines, the bladder, the skin, the eye – all can be damaged by chemical agents. Toxicity may be exerted by some chemicals on the developing embryo and fetus. It is convenient to categorize chemicals by the organ or system of the body that is the target for their toxicity, so we refer to liver toxicants, nervous system toxicants, dermal toxicants, and so on. Some chemicals will fall into only one of these categories, but most fall into several. Moreover, as exposure conditions change, so may targets.

This type of categorization, while convenient, might be misleading. It perhaps suggests that all chemicals having a common target produce the same type of toxic effect on that target. This is not the case, and we shall reveal several examples in the next chapter.

Chemicals causing certain adverse effects are singled out for special treatment. Those capable of producing excess tumors in any of many possible sites of the body are classified as *carcinogens*, and not according to the target on which they act (although they may be

---

[3] Physicians, of course, refer to diseases as chronic if they persist a long time in the patient, or if they have been a long time in development. This is a perfectly appropriate usage.

subcategorized as lung carcinogens, liver carcinogens, etc.). Chemicals causing birth defects of many different types are classified as *teratogens* (from the Greek "teras," meaning "monster"). Some chemicals alter behavior in undesirable ways, and so are classified as behavioral toxicants. These are a few of the special categories toxicologists have come to rely upon as they go about organizing their knowledge.

Some toxic effects are reversible. Everyone has been exposed to some agent, household ammonia for example, that produces irritation to the skin or eyes. Exposure ends and, sometimes perhaps with a delay, the irritation ends. Some readers have no doubt been poisoned on occasion by the ingestion of too much alcohol. The effects here also reverse. The time necessary for reversal can vary greatly depending upon the severity of the intoxication and certain physiological features of the person intoxicated. But most people also realize that chronic alcohol abuse can lead to a serious liver disorder, cirrhosis, which may not reverse even if alcohol intake ceases. This type of effect is irreversible or only very slowly reversible. It is important in making a toxicological evaluation to understand whether effects are reversible or irreversible, because one is obviously much more serious than the other.

# Risk

In the final analysis we are interested not in toxicity, but rather in *risk*. *By risk we mean the likelihood, or probability, that the toxic properties of a chemical will be produced in populations of individuals under their actual conditions of exposure.* To evaluate the risk of toxicity occurring for a specific chemical at least three types of information are required:

(1) The types of toxicity the chemical can produce (its targets and the forms of injury they incur).
(2) The conditions of exposure (dose and duration) under which the chemical's toxicity can be produced.
(3) The conditions (dose, duration) under which the population of people whose risk is being evaluated is or could be exposed to the chemical.

It is not sufficient to understand any one or two of these; no useful statement about risk can be made unless all three are understood. It matters not whether a chemical can produce severe forms of toxicity if no-one is exposed to it. And, it may matter a lot if huge numbers of people are exposed to a substance which appears to have very weak or no toxicity, when knowledge of its toxic properties is seriously deficient. It may take the next several chapters to create a thorough understanding of these matters, but they are the heart of the lesson of this book.

## Identifying the toxic properties of chemicals

It is time to inquire about the methods used to identify the toxic properties of chemicals. So far a few key principles have been introduced and some information on specific substances has been discussed, but little has been said about how these principles and information have been learned. Without some appreciation of the basic methods of toxicology, and what can and cannot be accomplished with them, it will not be possible to gain a solid understanding of the strengths and, more importantly, the limitations in our knowledge of chemical risk.

Toxic properties are identified in three basic ways: through *case-reports*; with the tools of *epidemiology*; and through *laboratory studies*, typically involving animals but also involving microorganisms, cells, and even parts of cells. Laboratory studies are of two types. The first involves what is called the *toxicity test* or *bioassay*, a study in which chemicals are administered in relatively standardized ways to groups of laboratory animals or other organisms, and observations are made on any adverse effects that ensue. The second type of laboratory study involves whole animals or parts of them (collections of cells, for example), and is designed to provide information on pharmacokinetics and on the mechanism of toxic action. Examples of mechanistic studies and how they contribute to an understanding of toxicity and chemical risk will be brought forward later in the book, in connection with discussion of specific agents, and in Chapter 9.

Case-reports are just what they sound like: reports, typically from physicians, regarding individuals who have suffered some adverse health effect or death following exposure to a chemical in their environment. Case-reports usually involve accidental poisonings, drug overdoses, or homicide or suicide attempts. They have been instrumental in providing early signals of the toxic properties of many chemicals, particularly regarding acute toxicity, and occasionally can be valuable indicators of the effects of chronic or subchronic exposure. Much of the very early information concerning the carcinogenic properties of arsenic, for example, came from physicians who observed unusual skin lesions and cancers in some of their patients treated with Fowler's solution, a once widely used arsenic-containing medicine.

Case-reports do not derive from controlled scientific investigations, but rather from careful and sometimes highly sophisticated scientific and medical detective work. Evidence is pieced together from whatever fragments of information are available, but is rarely definitive. Establishing causality, especially when effects are delayed and the exposure situation not clearly understood, is generally not possible. In the absence of controls, it is virtually impossible to know whether the patient would have contracted the disease or injury if exposure to the chemical had never occurred. It is also difficult in these situations to satisfy the toxicologist's goal of understanding the size of the dose necessary to produce toxicity. Learning just how much exposure took place as the result of an industrial accident is obviously very difficult. So while case reports will continue to provide clues to risks in the environment, they are perhaps the least valuable source of information for identifying toxic properties. They are typically referred to as "hypothesis generating." A series of what appear to be consistent case findings can generate a hypothesis about causality that can be tested in controlled studies.

Epidemiological studies are a far more important source of information on the effects of chemicals in humans. The epidemiologist tries to learn how specific diseases are distributed in various populations of individuals. Attempts are made to discover whether certain groups of people experiencing a common exposure situation (workers engaged in a common activity, for example, or patients taking the same medicine) also experience unusual rates of certain diseases. Epidemiologists may also try to learn whether groups of individuals having a disease in common also shared a specific type of exposure situation. Such studies are not strictly controlled, in the same way that laboratory studies are controlled. Epidemiologists attempt to take advantage of existing human exposure situations and, by imposing certain restrictions on how data from those situations are to be analyzed, seek to convert them to something approaching a controlled laboratory study. When this works, such studies can provide immensely valuable information about the toxic properties of chemicals in human beings; perhaps the most valuable.

Creating something approximating a controlled study out of a "natural" exposure situation is, it must be added, fraught with hazards, many of which simply cannot be overcome. It is rare that any single epidemiology study provides sufficiently definitive information to allow scientists to conclude that a cause–effect relationship exists between a chemical exposure and a human disease. Instead epidemiologists search for certain patterns. Does there seem to be a consistent association between the occurrence of excess rates of a certain condition (lung cancer, for example) and certain exposures (e.g., to cigarette smoke), in several epidemiology studies involving different populations of people? If a consistent pattern of association is seen, and other criteria are satisfied, causality can be established with reasonable certainty.

The difference between establishing that two events are "associated," in a statistical sense, and the difficult task of establishing that one event "causes" the other, will be discussed more fully in Chapter 6 "Identifying carcinogens," as will many other features of the epidemiology method.

The "gold standard" for human studies is called the randomized, controlled clinical trial. Such trials are close to experimental animal studies, but for obvious ethical reasons, they cannot be conducted to identify toxicity. They are, instead, designed to determine whether certain pharmaceutical or nutritional regimens, for example, reduce the risks of disease. They may provide information about adverse side effects, but they are not designed for studying toxicity.

We would, of course, prefer not to see anything but negative results from epidemiology studies. In an ideal world information on toxic properties would be collected before human exposure is allowed to take place, and that information would be used to place limits on the amount of human exposure that is permissible. If mechanisms existed to enforce those limits, then excess chemical risk would not occur and, it obviously follows, would not be detectable by the epidemiologist (unless, of course, the data or methods for setting limits were in error).

The world is, of course, not ideal, and over the past 100 years human exposures to thousands of commercially produced chemicals and the by-products of their production have been allowed to occur prior to the development of any toxicity data other than those related to short-term exposure. During the 1950s and 1960s various federal laws were enacted requiring the development of toxicity information prior to the marketing of certain classes of commercial products – food and color additives, pesticides, human drugs. The Toxic Substances Control Act (1976) imposed similar requirements on certain other classes of industrial chemicals. So in the past few decades we have begun to take steps towards that "ideal" world. The EU is currently attempting to move in giant steps towards that goal.

Epidemiology studies are, of course, useful only after human exposure has occurred. For certain classes of toxic agents, carcinogens being the most notable, exposure may have to take place for several decades before the effect, if it exists, is observable – some adverse effects, such as cancers, require many years to develop. The obvious point is that epidemiology studies cannot be used to identify toxic properties prior to the introduction of a chemical into commerce. This is one reason toxicologists were invented!

Another reason relates to the fact that, for methodological reasons, epidemiology studies cannot provide telling information in many exposure situations. It is frequently not possible to find a way to study certain situations in meaningful ways. Recognizing the limitations of the epidemiologic method, and concerned about ongoing human exposure to the many substances that had been introduced into commerce prior to their having been toxicologically well characterized, scientists began in the late 1920s to develop the laboratory animal as a surrogate for humans.

To learn about toxicity prior to the marketing of "new" chemicals, and to learn about the toxicity of "old" chemicals that did not have to pass the pre-market test: these are two of the major reasons toxicologists have turned to testing in laboratory animals.

In addition to the fact that animal tests can be applied to chemicals prior to marketing, such tests hold several advantages over epidemiology studies. First, and most important, is the fact that they can be completely controlled. We use this term in its scientific sense. Simply put, a toxicity study is controlled if the *only* difference between two groups of experimental animals is exposure to the chemical under study in one group and the absence of such exposure in the other. Only when studies are strictly controlled in this way can it be rigorously established that adverse effects occurring in one group and not in the other are *caused* by the agent under study. Experimental animals have uniform diets, come from almost identical genetic stock and are housed in well-controlled environments. Even the controlled clinical trial cannot match this experimental ideal.

Using laboratory animals also permits the toxicologist to acquire information on all the targets that may be adversely affected by a chemical, something that is not achievable using epidemiological science. Animals can be extensively examined by the toxicologist and the pathologist, whereas the epidemiologist is usually limited to whatever specific diseases are recorded for the population under study and for suitable "controls."

The obvious disadvantage of animal studies should at least be noted – laboratory animals are not *Homo sapiens*. Toxicologists use rats and mice, sometimes dogs, hamsters, guinea pigs, and even pigs – all mammals having the same basic biological features of humans. Much empirical evidence exists to show that laboratory animals and human beings respond similarly to chemical exposures. Dr. David Rall, former Director of the National Institute of Environmental Health Sciences, and some of his associates put the matter this way:

The adequacy of experimental data for identifying potential human health risks and, in particular, for estimating their probable magnitude has been the subject of scientific question and debate. Laboratory animals can and do differ from humans in a number of respects that may affect responses to hazardous exposures. . . . Nevertheless, experimental evidence to date certainly suggests that there are more similarities between laboratory animals and humans than there are differences. These similarities increase the probability that results observed in a laboratory setting will predict similar results in humans.

Toxicologists do not, however, have convincing evidence that every type of toxic response to a chemical observed in species of laboratory animals will also be expected to occur in similarly exposed human beings. To make matters more complicated, there are many examples of different species of laboratory animals exhibiting different responses to the same chemical exposure!

The "nuts and bolts" of animal testing, and the problems of test interpretation and extrapolation of results to human beings, comprise one of the central areas of controversy in the field of chemical risk assessment. They shall be with us, in one form or another, for the remainder of this book. Suffice it to say at this point that animal tests are extensively used to identify the toxic properties of chemicals – in part because animals can be good models for humans and in part because we do not have other good choices – and will continue to be used for that purpose for a long time to come. We shall now begin to show how this is done.

## Studying acute toxicity

Systematic investigation of the toxic properties of a chemical usually begins with identification of what is technically called the acute lethal dose–50 ($LD_{50}$): this is the (single) dose of a chemical that will, on average, cause death in 50% of a group of experimental animals.

The $LD_{50}$ has become one standard measure of a chemical's acute toxicity. It is obtained by administering a range of doses of the chemical of interest to several different groups of experimental animals – a bioassay is used. The objective is to expose each group of animals to a dose sufficiently high to cause a fraction of them to die. A typical result, for example, might see 9 of 10 rats die in the group receiving the highest dose, perhaps 6 of 10 in the next highest group, 3 of 10 in the next, and 1 of 10 in the group receiving the lowest dose. Doses are often administered using a stomach tube – the so-called *gavage* method – to allow the toxicologist an accurate quantitative measure of the delivered dose.

If this type of dose and response (in this case, the response is death) information is available, a simple statistical technique is applied to estimate the $LD_{50}$ – the dose that will on average cause death in 5 of 10 animals, or 50% of the animals in any similar group were the test to be repeated.

Note that it would be possible to administer a dose sufficiently high to kill all the animals in every test group; that dose and every imaginable higher dose represents the $LD_{100}$. But this type of information is not very useful to the toxicologist; the way in which lethality changes with dose and the point at which a single dose does not appear to have lethal potential are much more telling pieces of information.

Note also one other extremely important point; in the range of doses used, not every animal in each test group died. Only a certain fraction responded to the dose by dying, even though all animals in each test group are of the same species, sex, and even strain. Laboratory *strains* are members of the same species (e.g., rats), that are very closely related because of genetic breeding; biologists thus refer to the Wistar strain of rat, or the Sprague–Dawley strain, or the Fisher 344 strain, and so on. These animals have been bred to achieve certain characteristics that are desirable for laboratory work.

The $LD_{50}$, then, represents the dose at which animals have a 50% probability, or *risk*, of dying. This is our first specific example of risk information.

Identifying the $LD_{50}$ is not the only purpose of the acute toxicity study. The $LD_{50}$ provides a reasonably reliable indication of the relative acute toxicities of chemicals, and this is obviously important. But an even more important reason exists for conducting such tests, and that is to prepare the way for more extensive study of subchronic and chronic exposure.

During the acute toxicity determination the toxicologist carefully observes the animals for what are called "toxic signs."[4] If the animals appear to have difficulty in breathing, this sign indicates an effect of the chemical on the respiratory system. Tremors, convulsions, or hind limb weakness suggest the chemical damages the nervous system, or actually the neuromuscular system (nerves control muscle response). Redness or swelling of the skin points to a dermal toxicant, or perhaps a response of the immune system. These types of observation help to identify the specific targets that the particular chemical may affect in tests of longer-term duration, and the toxicologist can plan accordingly.

Knowledge of the range of doses that causes death also helps the toxicologist select doses for subchronic and chronic studies, which have to be conducted at doses that do not cause acute toxicity or early death.

There are some other acute toxicity tests in which non-lethal outcome are sought. These include studies of the amount of chemical needed to cause skin or eye irritation or more serious damage. Test systems developed by J. H. Draize and his associates at the Food and Drug Administration in the early 1940s were used to study ocular effects. Warning labels on consumer products were typically based on the outcome of the Draize test.

The Draize test in particular commanded much attention from animal rights activists, because it involved direct introduction of chemicals, typically consumer products of many types, into the eyes and onto the skin of rabbits. Ethical issues of several types arise in connection with the use of animals for toxicity testing. There are widely accepted guidelines concerning the appropriate care to be given to laboratory animals. These guidelines assume that animals can be ethically used for toxicity testing and other types of scientific and medical endeavors, but many in the animal rights movement and even some toxicologists question this premise, to greater or lesser degrees. These issues are important ones to toxicologists, and a significant segment of the toxicology community is now examining alternative means for acquiring some types of toxicity information. Some progress is being made in this important area, particularly regarding acute tests for dermal and ocular toxicity.

---

[4] Symptoms are what human patients can tell doctors about. An animal can't tell the toxicologist if it has a headache or an upset stomach. The toxicologist reads the "signs."

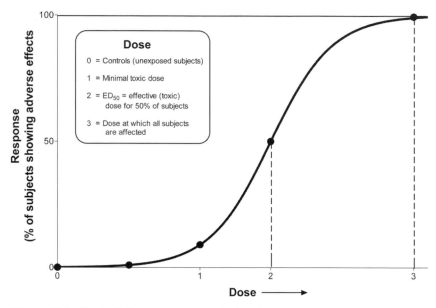

*Figure 3.1  Typical dose–response relationship.*

## Dose–response relationships

Figure 3.1 presents graphically what we have just described as the outcome of an acute toxicity study. What is seen is called a *dose–response* relationship. Doses (called 0 (control), 1, 2, and 3) are plotted against toxic response. The latter is expressed in the figure as the percentage of test animal subjects exhibiting a specific effect. If the effect is death, then dose 2 represents what we have called the $LD_{50}$. The response could be any manifestation of toxicity, and could result from dosing of any duration.

In many cases the percent or fraction of test subjects responding does not change; rather, all subjects respond at all non-zero doses and the *severity* of the effect increases. Reductions in body weight gains during a period of growth, relative to those of controls, are a frequent manifestation of toxicity, and one in which severity – the extent of such reductions – increases as dose increases.

With some possible exceptions (to be discussed in Chapter 9), dose–response relations identical or similar to those shown in Figure 3.1 are observed for all expressions of chemical toxicity. Indeed, the absence of such a dose–response relationship is often used as evidence that a chemical has not caused a particular response. Criteria for causation in

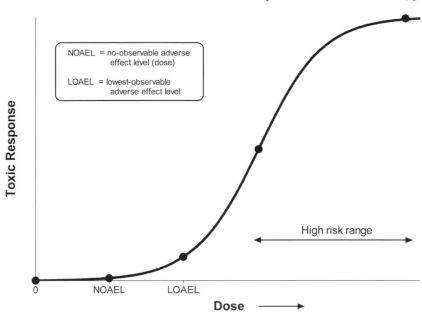

NOAEL = no-observable adverse
effect level (dose)

LOAEL = lowest-observable
adverse effect level

High risk range

NOAEL     LOAEL

**Dose** ⟶

*Figure 3.2 Dose–response relationship with description of NOAEL and LOAEL.*

epidemiology studies include consideration of the presence or absence of such relationships. (Not infrequently, for example, a low dose–response might be greater than that of the control. Such an observation suggests the chemical causes the particular response. But it is seen that no such increase occurs at higher dose. The absence of a dose–response relationship is evidence that the low dose "response" simply represents normal variation, and is not due to the chemical exposure. As we shall see, control animal responses for particular effects are rarely zero, and some can exhibit considerable variation.)

**NOAELs and LOAELs**

The no observed adverse effect level (NOAEL) is the highest dose, of the doses used in a study, that causes no adverse effect distinguishable from what is observed in control animals (Figure 3.2).

Note that the "true" no-effect dose may be greater than the NOAEL; the latter is in part an artifact of dose selection by the designers of the study. As the acronym suggests the LOAEL is the lowest dose at which some adverse effect is observable. Note also that "dose" could be any

*Calculated Risks*

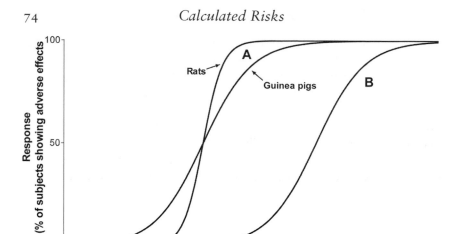

- Compound **A** is more toxic than compound **B**
- The response to compound **A** is more variable in guinea pigs than in rats
- The human population would be expected to exhibit still greater variability

*Figure 3.3  Comparative toxicity and variability in response.*

of several different measures (administered dose; level in inhaled air; blood level, etc.) as discussed in Chapter 2.

The subjects of NOAELs and LOAELs are critical to the risk assessment process, and we shall be referring to them throughout the book.

Figure 3.3 teaches some additional and important features of dose–response relationships. Such relationships are depicted for two different compounds (A and B), and responses in two different species, rats and guinea pigs, are shown for compound A. Because the dose–response relationship for compound B is to the right of that shown for A, we can conclude that B is less toxic than A, at least for the particular response plotted here (according to our principles, such a pattern could be reversed for some other manifestation of the toxicity of A and B). As seen in the figure, toxic responses to B consistently occur only at higher doses than they do for A, so B is less toxic.

The dose–response relations for compound A also appear to be somewhat different in rats and guinea pigs. In the case of rats there is an extremely sharp increase in response as dose increases. In guinea pigs, there is a slower rate of increase. This difference suggests that guinea pigs exhibit greater *variability* in response to A than do rats. Although the dose causing a 50% response rate is the same in the two species, the toxic response in guinea pigs begins at a lower dose than it does in rats, and increases more slowly – the difference between the LOAEL and the dose causing close to a 100% response is larger for guinea pigs than for rats. If we could develop such a dose–response relationship for humans, it would exhibit much greater variability in response than does the relationship for guinea pigs. Dealing with the problem of variability in response is one of the central issues in risk assessment, as we shall see in Chapter 8.

Dose–response relationships figure prominently in the development of risk assessments, and we shall have much to say about them.

## Subchronic and chronic tests

Consider some new chemical that might be a useful pesticide. The manufacturer is required by federal law to develop all the toxicity data necessary for an evaluation of its risk to human health, prior to its commercial introduction. If the pesticide were to be approved ("registered," in EPA parlance) people are likely to be exposed to residues of the pesticide in certain foods. Workers might be exposed during manufacture and product formulation, or during application of the pesticide, and perhaps even when harvesting treated crops. Exposure might occur for only a few weeks each year for the workers, but could be fairly regular for the general population for a large part of a lifetime, if the product is commercially successful. The EPA is responsible for specifying the toxicity tests to be performed, the appropriate design of these tests, and the controls that the testing laboratory needs to exercise to ensure the integrity and quality of the test data. The EPA must receive and evaluate the test data (along with a great deal more data concerning pesticide usage, residue levels, environmental fate, and toxicity to non-human organisms, also developed by the manufacturer), and then find health risks to be negligible, before the pesticide can be registered and sold.

The manufacturer's own toxicologists will begin testing with a determination of acute toxicity. If it appears the chemical is not unduly toxic, general subchronic and chronic, and a variety of specialized tests will be planned. A number of design issues need to be considered.

### Route of administration

The general population will be exposed by ingesting pesticide residues on certain foods, so most of the tests will involve administration of the chemical mixed into the diet of the test animals. Some studies in which the chemical is put into a vapor or aerosol form will be conducted, because some pesticide workers may inhale the substance. Skin toxicity studies are also needed because of possible pesticide worker exposure by this route. The principle here is straightforward: try to match the routes of exposure to be used in the tests to the likely routes of human exposure.

### Test species

Most tests, both general and specialized, will be conducted using rats and mice. Dogs are used for certain specialized studies and rabbits for others. These species, specifically certain strains of them, have a long history of use as test subjects; their behavior in laboratory settings is understood and their dietary and other needs are well characterized. Another important consideration in the selection of test species and strains is knowledge of the types of diseases common to them. In the normal course of their lives certain diseases will "spontaneously" develop in all species, so some of the animals assigned to the untreated, control groups in the toxicity studies will naturally develop certain disease conditions. (Many toxicologists use the unfortunate term "spontaneous" when referring to diseases of unknown cause.) A classic example of this is the development of certain kidney diseases in elderly male rats. Knowledge of the normal range of "background" disease rates, some of which are highly variable, in untreated animals is important to help toxicologists understand whether observed effects are chemically induced or normal.

An issue of obvious importance in test species selection is the degree to which test results can be reliably applied to human beings. As we noted in the last chapter this is one of the principal problems in the evaluation of human risk, and we shall get back to it in the later chapters on risk assessment. For now, emphasis is on the selection of animal species and strains for their known reliability as experimental subjects. To put it in stark (but honest) terms – the animals are used as toxicity measuring devices.

### Controls

No study design is acceptable unless appropriate control animals are used. These are animals of the very same species, sex, age, and state of health as those to be dosed with the test chemical. The only difference between the controls and treated animals is the absence in the control group of exposure to the chemical to be studied. Animals are assigned to test and control groups in a way, called randomization, to reduce the potential for biased outcomes.

### Number of test subjects

If a pesticide is registered by the EPA, millions of people might become exposed to it. Obviously it is not possible to use millions of laboratory animals in a test, and even thousands will present a logistical nightmare. For practical reasons, most tests are performed with 20 to 60 animals of each sex in each of the dose groups (a "dose group" is a group of animals all of whom receive the same daily dose). Tests involving these numbers of subjects are obviously limited in some ways, and the toxicologist needs to consider these limits during the ultimate risk evaluation. But let us go into this matter now, because it is exceedingly important.

Suppose that, *unknown to the toxicologist*, a certain dose of a chemical causes a serious toxic effect – damage to certain brain cells – in one of every 50 exposed subjects. In other words, there is a 2% risk of this form of toxicity occurring at our specified dose. Suppose this same dose is administered to a group of 50 rats, and the examining pathologist sees that one animal develops this particular form of brain damage. He also notes that none of the 50 untreated control animals develops the problem. Is it correct to conclude that the chemical caused this effect? The toxicologist finds a friendly statistician (they do exist) and is informed such a conclusion cannot be reached! Why not?

The statistician's role is to determine whether a disease rate of 1 in 50 is truly distinguishable from a disease rate of 0 in 50. The statistician will point out that there is only a very small chance (and chance, or probability, is what the statistician calculates) that this observed difference between the two groups of animals is actually due to the presence of the chemical in the diet of one of the groups, and its absence in the other. In fact, the statistician will state that not until the difference in disease rate is 0/50 versus 5/50 is there reasonable probability that the observed difference is actually due to the chemical. In other words, a difference in disease rate of at least 10% (0/50 versus 5/50) is necessary to achieve what is called a *statistically significant* effect. The difference necessary to achieve statistical significance will be smaller if more animals are used in the test (larger denominator), and larger if fewer animals are used.

Now it has been stated that, unknown to the toxicologist, the given dose of this chemical actually does cause a 2% increase in the rate of occurrence of this brain lesion. The problem is that, in an experiment involving 50 animals in a test group, the toxicologist cannot call an observed rate of 2% a true effect with any statistical legitimacy; indeed, not until the rate reaches an excess of about 10% can they conclude on statistical grounds that there is a difference between the responses of the two groups of animals. We are talking here about *what we can claim to know* with reasonable certainty as a result of experiments with limited numbers of subjects. If the number of animals in a test group is increased, lower disease rates can be detected, but rates very much below 5–10% cannot be achieved with groups of practical size.

What all this means – and this is of much concern in risk assessment – is that the animal tests we are describing cannot be used to detect excess diseases occurring with frequencies below 5–10%, and these are fairly large risks, well above what we would deliberately tolerate in most circumstances (although pack-a-day smokers tolerate lifetime cancer risks about this high for themselves). The 2% excess risk in our example is also fairly large, but could not be detected in our experiment – it is a real risk, but because of inherent limits on what we can claim to know based on these types of test, it remains hidden from us. The risk assessor has a way of dealing with this type of limitation, and it shall be discussed in Chapters 7 and 8. For now, we simply restate that the numbers of animals assigned to toxicity test groups are largely determined by practical considerations, and that interpretation of test results needs always to consider the limitations imposed by the use of relatively small numbers of animals.[5]

The result from our hypothetical experiment is sometimes referred to as a "false negative"; we erroneously conclude there is no effect when in fact there is one. Of course we can know our conclusion is false only if we somehow develop evidence that there is a real adverse effect, through additional study. Toxicologists also worry about "false positives," but for adverse effects it would seem that "false negatives"

---

[5] Cost is also an issue. Currently, a chronic feeding study involving two sexes of two rodent species can cost more than 1 000 000 dollars. Inhalation studies are more expensive. Many studies in addition to chronic studies may be required.

are of greater concern. We cannot eliminate such false outcomes; but good experimentation can reduce their importance.

Before this topic is left behind, it should be noted that statistical significance is by no means the only consideration in interpretation of toxicity test results. If, in our particular case, the pathologist were to inform us that the brain lesion observed was extremely unusual or rare, we should certainly hesitate to dismiss our concerns because of lack of statistical significance. The toxicologist needs equally to understand "biological significance," and, in this case, would almost certainly pursue other lines of investigation (perhaps an ADME study to determine if the pesticide reaches the brain, or a toxicity test in other species) to determine whether the effect was truly caused by the chemical.

### Dose selection

The usual object of test dose selection is to pick, at one extreme, a dose sufficiently high to produce serious adverse effects without causing early death of the animals, and, at the other, one that should produce minimal or, ideally, no observable adverse effect – a NOAEL. At least one and ideally several doses between these extremes are also selected. We wish to come out of the experiment with a dose–response curve such as that depicted in Figure 3.2.

Some sophisticated guessing goes into dose selection. Knowledge of the minimum acutely toxic dose helps the toxicologist pick the highest dose to be used; it will be somewhere below the minimum lethal dose. There is usually little basis for deciding the lowest dose; it is often set at some small fraction of the high dose. Whether it turns out to be a NOAEL will not be known until the experiment is completed. Sometimes bioassays have to be repeated to identify the NOAEL.

### Duration

The toxicologist usually moves from studies of a single exposure to ones in which animals are exposed on each of 90 consecutive days. The 90-day subchronic study has become a convention in the field. Rodents usually live 2–3 years in the laboratory, so 90 days is about 10% of a lifetime. An enormous amount of 90-day rodent toxicity data have been collected over the past several decades and have played key roles in judging the risks of environmental chemicals.

Our pesticide, we have said, will end up in certain foods and
people could ingest residues over a portion of their average 70–75 year
lifespan much greater than 10%. So the toxicologist needs also to
understand the toxicity associated with chronic exposure, usually set
at about 2 years in rats and mice. The subchronic toxicity results are
usually used to help plan the chronic study: to identify the doses to
be used (usually a range of doses lower than those used subchron-
ically) and any unusual forms of toxicity that need to be examined
with special care.

One of the toxic effects the chronic study is designed to detect is
cancer formation. Some toxicologists believe, in fact, that cancer is
the only form of toxicity not detectable in 90-day studies! Indeed, it
is difficult to find many examples of forms of toxicity occurring in
chronic studies that were not detectable, at higher doses, in 90-day
studies. It appears that, in most cases, the chronic exposure allows the
effects that were detected in 90-day studies to be detected at lower
doses, but does not reveal new forms of toxicity, except possibly cancer. This is not a sufficiently well-established generalization to support
rejection of the need for chronic studies, and, of course, the toxicologist obviously needs to determine whether a chemical can increase the
rate of tumor formation. So chronic studies will be around for some
time.

In Chapter 6 we pursue in greater detail the problem of identifying
carcinogens using experiments.

**Observations to be made**

The extent, frequency, and intensity of observations to be made on
the treated and control animals vary somewhat among types of study.
In general, the toxicologist monitors at least the following parameters
during the study: survival pattern, body weight, food consumption
rate, behavior patterns, and blood and urine chemistry.

The animals receive a battery of clinical measurements, much like
those people receive when they leave samples of blood and urine for
testing after a medical examination. It turns out that body weight –
reduced weight gain for growing animals or weight loss for adults –
is a particularly sensitive indicator of toxicity. Its measurement does
not provide much of a clue about the nature of the toxic effect that
is occurring, but it is considered an adverse response in and of itself.
In some cases it is due to reduced food consumption (and this is why
food consumption is measured carefully), because the addition of the

chemical to the diet makes the food unpalatable. In such a case, the chemical is obviously not producing a toxic effect; such a finding simply means that the experiment has to be repeated in a way that avoids the problem – in most cases this means undertaking the tricky task of introducing the required dose into each animal by stomach tube; the gavage method of dosing.

At the end of the study animals remaining alive will be killed, and examined by a pathologist. So will any animals that die during the course of the study, assuming their deaths are discovered before their tissues have begun to decompose.

The pathologist will first visually examine each animal inside and out. About 40 different tissues and organs will be taken and prepared for examination under a microscope – the so-called histopathological examination. As in the case of the pathologist who looks at tissues from people, the animal pathologist is characterizing the disease state or type of injury, if any, to be found in particular tissues. The pathologist does this "blind" to the source of the tissue, that is, without knowledge of whether the tissue came from the treated animals or the control animals.

So the animals used in the toxicity test, both treated and control animals, are subjected to extremely thorough "medical monitoring," even to the point of sacrificing their lives so that the toxicologist can learn in minutest detail whether any of their tissues have been damaged. Obviously, there is no way such thorough information could ever be collected from any imaginable study of humans exposed to chemical substances.

## Conducting the bioassay

Once the design is established, a protocol will be prepared. All the critical design features and the types of observations to be made, and even the statistical methods to be used to analyze results, are specified in advance. Toxicologists, chemists, pathologists, and statisticians are typically involved in drafting the protocol.

Some aspects of the mechanics of testing deserve mention. First, animals to be put into the control group and to the groups to be treated with the chemical need to be assigned in a completely random fashion. The animals are usually selected for testing not long after weaning, while they are still in a growing phase, and care must be taken to avoid any discrimination among the groups with respect to factors such as weight – the person assigning the animals to various groups should have a procedure to allow completely random selection of animals.

A second factor concerns the purity of the diet and water received by the animals. Careful chemical analysis is needed to ensure the absence of significant amounts of highly toxic chemicals, such as aflatoxin, metals such as lead, arsenic, or cadmium, or certain pesticides, that may be present in water and various feed ingredients.

If the oral route is to be used, the chemical may be mixed with the diet, dissolved in drinking water, or delivered by a tube to the stomach (gavage). An inhalation exposure requires special equipment to create the desired concentration of the chemical in the air to be breathed by the animal. In any case, the analytical chemist must be called on to measure the amount of the chemical in these various media after it has been added to guarantee that the dose is known with accuracy. Some chemicals decompose relatively quickly, or errors are made in weighing or mixing the chemical to achieve the desired diet, water, or air concentrations, so chemical analysis of these media is essential throughout the study.

There are many other features of toxicity studies that require careful monitoring and record-keeping, but they won't be mentioned here. Suffice it to point out that conducting a chronic toxicity study requires extremely careful control and monitoring. Indeed, a series of discoveries during the 1970s of poor record-keeping, sloppy animal handling and, in a few cases, deliberate recording of false information in study reports led to the promulgation of federal regulations concerning "Good Laboratory Practice." The regulations specify the type of data collection and record-keeping and additional study controls that must be documented for studies whose results will be submitted to federal regulatory agencies. It is foolhardy these days to conduct toxicity tests in laboratories that cannot demonstrate strict adherence to GLPs (note we are referring to toxicity tests, as distinct from toxicity research).

### Protocol standardization

Regulators have found it useful, indeed necessary, to standardize protocols for toxicity testing. International "harmonization" of toxicity testing protocols has, in fact, been promoted by governments, advocacy groups, and affected industries, as a major step towards reducing the need for repeat testing of the same chemical because one country's protocol guidelines contain somewhat different requirements than another's. Several websites listed in the *"Sources and recommended reading"* section can lead interested readers into the heart of the current, worldwide toxicity testing enterprise.

### Study evaluation

At the end of the toxicity test the toxicologists and pathologists list all the observations made for each individual animal in each dose group. Analysis of these results is needed to identify effects caused by the chemical under study. All observations – body weights, clinical measurements, histopathology, and so on – have to be included in the evaluation.

Two types of analysis are needed: one concerns the biological significance of the results and the other their statistical significance.

The statistician is interested in determining the chance, or probability, that the rates of occurrence of certain injuries in specific tissues are different from the rates of their occurrence in the same tissues of control animals. Observations of numerical differences are not sufficient to conclude that the difference is due to the chemical treatment and not to simple chance. The statistician has several well-established techniques to estimate the probability that an observed numerical difference in the occurrence of a particular effect between a group of control animals and a group of treated animals is due simply to chance. It is conventional in many areas of biological science, including toxicology, to conclude that an observed difference is a "real" one if there is less than a 1 in 20 probability that chance is involved. This is a scientific convention, not a law of nature. More or less strict criteria for statistical significance can be applied: the less the probability that we have observed a chance occurrence, the greater the probability that the chemical was responsible for that occurrence.

So the statistician does this type of analysis for each treatment group versus the control. When this is done the toxicologist can determine whether any of the observed effects are "statistically significant" – whether and to what degree of confidence the observed effect can be said to have been caused by the chemical treatment.

The toxicologist looks not only at the rates of occurrence of various adverse effects, but also examines the question of whether the severity of certain forms of injury is significantly greater in treated animals. "Severity" is not always quantifiable – it is in the eye of the pathologist. So some judgment beyond what the statistician can offer through an objective analysis is always necessary to complete an evaluation.

Statistics do not tell the whole story. Not infrequently effects are seen at very low frequencies (not at statistically significant rates) that the toxicologist may think important and likely to be due to the chemical. The typical case involves the appearance of very rare or highly unusual diseases or forms of injury in treated animals – diseases or injuries that historically have never been observed, or observed only at extremely low frequencies, in untreated animals.

Test interpretation is often made difficult because diseases occur with very high frequency in specific organs of untreated animals. If an average of 80% of aged, untreated, male rats of a certain strain normally suffer from kidney disease, then it becomes difficult to determine, both statistically and biologically, whether any observed increase in this same disease in treated animals is truly related to the chemical. The high background rate of the disease obscures the effect of the test chemical. The same type of problem plagues the epidemiologist. Searching for specific causes of diseases such as breast, lung, or colon cancers, that occur with relatively high frequency in human populations, is extremely difficult.

In the end the specific toxic effects that can be attributed to the chemical with reasonable confidence can be isolated from those that cannot be so attributed. As in most areas of science, there will almost always be effects falling into an ambiguous zone, or there will be effects that are of uncertain significance to the health of the animal. In the latter category are, for example, minor changes in the rate of occurrence of certain normal biochemical processes, typically at the lowest test dose, unaccompanied by any other sign of disease or injury. The toxicologist simply does not know whether such a change – which is clearly caused by the chemical treatment – has any adverse consequences for the health of the test animal. In fact, in recent years, regulatory attention has turned away from reliance on NOELs (the dose in which no biological change of any type is observed), and instead is focused on identification of the "no observed *adverse* effect level" (NOAEL).

## Specialized tests

So far the toxicologist has not made inquiries regarding a number of potentially important questions. Can the chemical harm a developing embryo and fetus, perhaps producing birth defects? Can the chemical reduce male or female fertility, or otherwise impair reproduction? Can the chemical injure the immune system, or alter behavior? Can it cause cancer or mutations?

These sorts of questions cannot be thoroughly explored with the general tests discussed so far. These tests do not, for example, provide for any mating of the animals. They do not allow chemical exposure of the females when they are pregnant. The chronic study may pick up an excess of tumors, but sometimes special chronic tests – called *cancer bioassays* – are performed to test the carcinogenicity hypothesis (Chapter 6). Studying the effects of chemicals on behavior, on certain neurological parameters, on the immune system, and on the materials of inheritance, requires test designs and measurement techniques that are different from those needed for generalized testing, and which cannot readily be built into the general tests. Moreover, many specialized tests are still in the developmental stage, and regulatory agencies tend to be reluctant to require their use until they have undergone validation of some type.

Typically, ADME studies are included in the battery of tests used to characterize the toxicity of chemicals, as well as other studies designed to trace the underlying molecular and cellular events that lead to toxicity. These studies of toxic mechanisms take many forms, and are better viewed as research studies; no general characterization of them will be made here, but some of the things such studies can reveal to aid understanding of risk will be mentioned at appropriate places in the remaining sections of the book.

## Overview of toxic mechanisms

The production of adverse effects by a chemical can be thought of as governed by two underlying processes. Pharmacokinetic processes govern the movement of a chemical from the external environment through the body, and include the production of metabolites (ADME, as described in the last chapter). If there is a sufficient amount of the chemical or, more often, one or more of its metabolites, at a site in the body where it can initiate some type of damaging event, and it is present for a sufficient amount of time, some form of toxicity can result. Pharmacodynamics is the name given to such *target site interactions*. When toxicity is measured, for example in some type of animal study, what is typically observed is a relationship between administration of the chemical and resulting toxicity. Pharmacokinetics and pharmacodynamics describe at the molecular, cellular, and

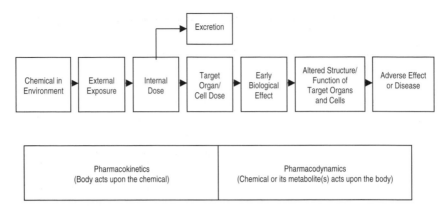

*Figure 3.4 From exposure to adverse effect or disease, advances in epidemiology and toxicology are bringing greater knowledge about each event in the process and about the fine molecular details of each.*

subcellular levels, the events underlying the production of toxicity (Figure 3.4).

Special tools are needed to study those underlying processes, and a significant fraction of scientists in the toxicology community are involved in such research. Sometimes the phrase "mechanism of toxic action" is used to describe these various underlying processes, although it is often used to describe only the pharmacodynamic piece of the picture. It is extraordinarily difficult to uncover all relevant mechanistic processes, but significant pieces of the puzzle of toxicity are known for many substances.

Before we embark on a descriptive survey of toxic and carcinogenic phenomena, the subjects of the next three chapters, it will be useful to provide a broad outline of the ways in which toxic injuries can be produced, and the ways in which they manifest themselves. This type of discussion will help to place into a unified context the descriptive material to come, and it will also aid in understanding concepts of toxic mechanisms as they relate to understanding human risk. The topic is presented in broad outline only; details are far more complex than is suggested here.

When the terms "toxic agent" and "toxicant" are used in the following, we refer to whatever chemical entity, the administered chemical (parent chemical) or one or more of its metabolites, is initiating or promoting damage. Also, it should be kept in mind that the same toxicant can, under different exposure conditions or in different species, produce different types of injury, by different mechanisms.

The discussion to follow classifies *mechanisms of injury* as occurring directly to cells (intracellular mechanisms), or as interfering with processes outside of cells that may indirectly bring about cellular damage (extracellular mechanisms). It will also be seen that, depending upon the mechanism of injury, cells *respond* in different ways. Injured cells may *degenerate* in different ways, or they may actually *proliferate*. The body also has mechanisms in place for the repair of damaged or dying cells, through processes usually described as leading to *inflammation*. Most of the following concerns events that occur once a toxic agent reaches its target; we have already discussed in some detail, in Chapter 2, the subject of the pharmacokinetic processes governing the movement of toxicants to those targets. Specific examples of toxicants acting by the various mechanisms described here will be offered in Chapter 4.

## Mechanisms of injury

### Intracellular mechanisms

Cytotoxicity is the general term used to describe toxicity at the level of the cell. It can be brought about in many ways, usually by a chemical interaction between the toxic agent and one or more components of the cell. Interactions can be permanently damaging or may lead to temporary injury that the cell is capable of repairing. Perhaps the most important sites of intracellular injury are *cell membranes*, the *cell nucleus* (home of DNA), *mitochondria* (home of energy production), and *endoplasmic reticulum* (home of the biosynthesis of the all-important protein molecules, essential for cell structure and, as enzymes, for the catalysis of all cellular reactions and for the metabolism of foreign chemicals).

Membranes separate cells from their external environment, and the internal components of cells from each other. Many biochemical processes taking place within cells occur on a framework of membranes. Toxicant interactions with membranes figure prominently in many types of toxic effect.

If a toxicant enters a cell it may interact with DNA or with any of the biochemical pathways necessary for the successful synthesis of this all-important molecule. Toxic agents may also interfere with the critical functions of DNA, so that genetic information encoded within it is not successfully translated into the synthesis of proteins. Direct toxicant damage to DNA may lead to mutations with many adverse consequences, among them the creation of permanently damaged daughter cells that continue to reproduce and lead to the abnormal growths we call benign and malignant tumors.

Any interference with protein synthesis, through alteration of DNA function, as just mentioned, or by damage to the structures called endoplasmic reticulum, the site of such synthesis, can be devastating in many ways, because proteins are not only essential for the many structures of cells, but also because they are the body's catalysts (enzymes) for all its essential biochemical processes.

Finally, we cannot neglect damage to the cell's energy center, the mitochondria. These organelles, if damaged, can cause a cell to shut down or become aberrant in many ways.

These various chemical interactions, whose consequences we shall describe below, usually involve some type of reaction between toxicant and one or more of the molecules that make up cells. The toxicant typically has within its chemical structure several collections of chemically reactive atoms, called functional groups, that initiate the injurious reactions. The reactive functional groups may be present in the parent chemical's structure or may be created during Phase I metabolism of the parent chemical. In some cases relatively inert chemicals may, simply because of their physical properties (e.g., a high propensity to dissolve fats) bring about cell damage. Again, the nature and extent of damage depends upon the concentration of toxicant at target sites, its residence time there prior to its excretion, and the nature of any interaction that ensues.

### Extracellular mechanisms

Like all of us, cells cannot survive without an external environment that operates successfully. The external environment brings oxygen and the nutrients needed to satisfy the cell's metabolic requirements. (Of course, they also bring foreign and potentially injurious chemicals to the cell.) These extracellular environments also allow for cells to maintain appropriate levels of fluids and electrolytes, and serve to remove cellular products that need to be excreted from the body. It is not hard to imagine how toxicant impairments of extracellular environments can be injurious.

When discussing extracellular environments we need to include the so-called regulatory systems of the body: the nervous system, the endocrine system, and the immune system. These systems of cells communicate with each other, often through hormones or other molecular entities, in highly complex ways. Toxicant damage to any component of these systems, or interference with the extracellular process by which communications occur, can affect many functions of the body, even many that are not components of these systems. They regulate huge numbers of processes in the body, and probably many more than we now understand. Systems toxicity can be toxicity of the most serious kind. The study of so-called "endocrine disruptors," chemicals that adversely alter one or more aspects of endocrine function, has recently attracted enormous interest throughout the world (more on this topic in Chapter 9).

## Responses to injury

### Degenerative responses

Degeneration refers to one type of cellular response to the injurious effects described in the preceding section; it is often observable at the level of the organ or tissue comprised of these cells. Reduction of cell size or growth rate (the latter called *atrophy*), excessive accumulation of water, fat or other cellular material, degradation of cellular structures, and cell death (*necrosis*), are all manifestations of degenerative responses. Toxicologists who specialize in pathology can make these diagnoses. Some are obvious with simple gross observation of tissues, while others can be seen only under microscopic examination.

Degenerative changes can result from all types of intra- and extracellular injuries. The most serious, necrosis, can threaten the life of the entire organism if it is extensive in any organ or tissue of the body. Pathologists use the suffixes "osis" or "opathy" when describing a degenerative change.

### Proliferative responses

A cellular response to either intra- or extracellular damage that appears to be opposite to that of degeneration is called proliferation (suffixes "plasia" or "oma"). Proliferative changes generally refer to increased growth of cells or cellular components. *Hypertrophy* refers to increase in cell and organ size; *hyperplasia* refers to increase in cell numbers. At one extreme, proliferation may represent little more than an adaptive response to a minor level of chemical injury; the latter is common in the liver, where the presence of a foreign chemical induces changes in liver cells that are working hard to eliminate the stranger through increased production of P-450 enzymes. At the other end of the range of proliferative responses are those of cells that have sustained serious injury in the form of genetic damage. In these cases, hyperplasia (rapid increase in cell numbers) may turn into *neoplasia* – rapid proliferation of "new" cells, or cancers. We shall take a close look at neoplasia in Chapter 5.

### Inflammatory responses

The third type of response to injury, which falls under the general heading of inflammation (suffix "itis"), manifests itself in several complex ways. It involves extracellular processes and cells of the immune system. Inflammation is often part of the road to repair from injury, but the inflammatory process can, if extensive, be highly damaging. Inflammation can be acute or chronic in nature. Repair can occur by *regeneration* of cells, for example by enhanced growth of adjacent cells; or it can occur by a process called *fibrosis*. Some examples of inflammatory responses and repair are brought out in Chapter 4.

These somewhat simplified descriptions of mechanisms that initiate cellular injury, and of the ways in which cells and tissues respond to these injuries will, as noted at the outset, be helpful as we describe various manifestations of toxicity and carcinogenicity. We distinguish between "toxic injuries," which are typically seen in animal experiments and are usually described in the terms defined in the foregoing, and the various medical conditions we call "diseases." Many toxic responses can lead to disease, but we also consider toxic injuries to be adverse effects, whether or not they are known to lead to specific diseases.

In the next three chapters we describe some of the many ways chemical exposure can lead to toxic injury and disease, and then enter the final third of the book, devoted to the problem of "calculating" risks.

# 4

# Toxic agents and their targets

A complete treatment of the subject of this chapter would require many volumes. There are some toxicological data available on over 20 000 industrial chemicals, including pharmaceuticals and consumer products of all types, although for a very large proportion of these the data reflect only relatively short-term, high-dose exposures. The number for which both comprehensive animal and epidemiological studies are available probably does not exceed a few hundred. The rest have some intermediate degree of toxicological characterization. The data base is growing all the time and new regulatory initiatives, especially in the EU, are intended to accelerate that growth in the near future.

Numerous compendia are available and can be consulted if there is a need to acquire comprehensive knowledge on specific substances (see *Sources and recommended reading*). The intention of this chapter is not to provide anything even remotely complete about any given chemical. It is instead simply to illustrate with concrete examples the many toxicological principles and concepts we have been discussing, and to show the diverse ways chemicals of many different types can bring about harm, and the conditions under which they do so. We cover most of the significant targets of toxicity, provide a little background on the biological characteristics of those targets – their structure and functioning – and then show the several ways excessive chemical exposure can cause harm.

The specific type of harm we call cancer is left to later chapters, because it is so important and because there are so many aspects of cancer initiation and development that are unique. It would be a mistake to interpret the separate treatment of cancer as an indication that it is a far more important toxicological issue than those discussed in this chapter. Cancer as a disease phenomenon is obviously of enormous importance, but it may well be that the types of chemicals that are the central subjects of this book are less important causes or promoters of the cancer process than are those associated with various "lifestyle" choices (which are still "environmental" but which entail sources of risk that include large elements of personal choice and that are otherwise distinct from those on which we are focusing). We are learning that the types of substances that are the principal subjects of this book may be more significant contributors to other manifestations of human disease, namely those covered in this chapter.

We begin with the killers – toxic agents of many types that have the capacity to cause serious injury or death at relatively low dose delivered over relatively brief exposure periods. These are what we rightly call "poisons." We then turn to the "slow poisons," the many substances that produce their most serious effects when delivered over long periods of time, at doses well below those that are immediately dangerous. When we discuss the "slow poisons" we do so by categorizing them by their "targets": respiratory toxicants, liver toxicants, substances that damage the nervous or reproductive systems, and so on.

There are other ways to categorize toxicants. Many texts categorize them by chemical class (the metals, aldehydes and ketones, aromatic hydrocarbons, and so on). You will also find toxicants categorized by use or source (food additives, pesticides, air pollutants, cosmetics ingredients), and even by the mechanistic phenomena that cause their toxicity (metabolic poisons, DNA-damaging agents, cholinesterase inhibitors). Each of these various categories has value in the appropriate context, and the approach taken here is, I believe, the most useful for the ultimate use to which we shall be putting the knowledge: the conduct of toxicological risk assessments.

## Poisons from nature

Botulinum toxins are a collection of protein molecules that are extraordinarily poisonous to the nervous system. These toxins[1] are metabolic

---

[1] The name "toxin" is correctly applied to naturally occurring protein molecules that produce serious toxicity. There has been a tendency to broaden the use of this term to include other categories of toxic agents. We shall adhere to the proper usage in this book.

products of a common soil bacterium, *Clostridium botulinum*, which is frequently found on raw agricultural products. Fortunately, the bacterium produces its deadly toxins only under certain rather restricted conditions, and if foods are processed properly so that these conditions are not created, the toxins can be avoided. Food processors have to be extremely careful with certain categories of food – canned foods having low acidity, for example – because the slightest contamination can be deadly. In 1971 an individual succumbed after consuming a can of vichyssoise made by the Bon Vivant soup company. A massive recall of canned soups resulted. The Bon Vivant company vanished soon after this event; botulinum toxin not only killed the customer, but also exterminated the company.

Botulinum toxins can be *lethal* at a single (acute) dose in the range of 0.000 01 mg/kg b.w.! This amount of toxin is not visible to the naked eye: about a million lethal doses per gram of toxin! The initial symptoms of botulism typically appear 12–36 hours after exposure and include nausea, vomiting, and diarrhea. Symptoms indicating an attack on the nervous system include blurred vision, weakness of facial muscles, and difficulty with speech. If the dose is sufficient (and a very, very small dose can be), the toxicity progresses to paralysis of the muscles controlling breathing – the diaphragm. Death from botulism thus comes about because of respiratory failure. Botulinum toxins are regarded as the most acutely toxic of all poisons.

To illustrate the famous Paracelsusan claim that the dose distinguishes a drug from a poison, we can point to botox, an extremely dilute form of botulinum toxin that effectively softens skin wrinkles and also, at least temporarily, eliminates the spastic muscle contractions associated with conditions such as cerebral palsy. In 1965 at the age of 37, the famed concert pianist, Leon Fleisher, lost the use of his right hand. His condition, called focal dystonia, involves a type of neurological "misfiring" that causes some muscles to contract uncontrollably (in Fleisher's case those in certain fingers of his right hand). The pianist, after a few years of botox injections (which he has to repeat every six months), recently released his first recording since the 1960s in which he plays pieces requiring two hands (he found a living teaching and playing pieces written for the left hand).

Sucrose, which we all know as table sugar, can also be acutely toxic. I cannot locate any evidence of humans being killed by a dose of table sugar, but toxicologists can force enough into rats to cause death. A lethal dose of sucrose in rats is in the range of 20 000 mg/kg b.w. That is about as "non-toxic" as chemicals get to be. If humans are equally

Table 4.1 *Conventional rating scheme for lethal doses in humans*

| | Probable lethal oral dose for humans | |
|---|---|---|
| Toxicity rating | Dose (mg/kg b.w.) | For average adult |
| 1 Practically non-toxic | more than 15 000 | More than 1 quart |
| 2 Slightly toxic | 5000–15 000 | 1 pint–1 quart |
| 3 Moderately toxic | 500–5000 | 1 ounce–1 pint |
| 4 Very toxic | 50–500 | 1 teaspoon–1 ounce |
| 5 Extremely toxic | 5–50 | 7 drops–1 teaspoon |
| 6 Supertoxic | less than 5 | less than 7 drops |

sensitive, one would have to eat more than 3 pounds of sugar at one time for it to be lethal.

The two substances – sucrose and botulinum toxins – differ in lethality by about 10 billion times! The acute lethal doses of most chemicals fall into a much narrower range, but there are many substances near the two extremes of this distribution of lethal doses.

Clinical toxicologists have found it convenient to rate chemicals according to their potential to produce death after a single dose. The conventional rating scheme is as shown in Table 4.1.

Clearly, if a person has to ingest a pint or more of a chemical before his life is seriously threatened, this chemical is not a likely candidate for use in homicide or suicide, and is highly unlikely to be ingested accidentally in dangerous amounts. Chemicals rated in categories 5 or 6, however, need to be extremely carefully controlled.

Keep in mind that the rating chart presented above concerns only *acute, lethal* doses, received by the oral route. It only provides a very limited picture of the toxic properties of chemical agents, and should never be used as the sole basis for categorizing chemicals. Some chemicals that are "supertoxic" by the above rating have no known detrimental effects when they are administered at sublethal doses over long periods of time, while others in the same category and in lower categories do produce serious forms of toxicity after repeated dosing.

Some naturally occurring "extremely toxic" and "supertoxic" chemicals are listed in Table 4.2, along with their environmental sources and toxicity targets. Some of these are toxins found in the venom of poisonous snakes or in the tissues of certain species of

Table 4.2 *Some supertoxic chemicals of natural origin[a]*

| Chemical | Source | Principal toxicity target |
| --- | --- | --- |
| Botulinum toxin | Bacterium | Nervous system |
| Tetrodotoxin | Puffer fish (fugu) | Nervous system |
| *Crotalus* venom | Rattle snake | Blood/nervous system |
| Naja naja | Cobra | Nervous system/heart |
| Batrachotoxin | South American frog | Cardiovascular system |
| Stingray venom | Stingray | Nervous system |
| Widow spider venom | Black widow | Nervous system |
| Strychnine | Nux vomica[b] | Nervous system |
| Nicotine | Tobacco plant | Nervous system |

[a] Some of these chemicals, particularly the venoms, are protein or protein-like compounds that are deactivated in the gastrointestinal tract; they are poisonous only when injected directly into the blood stream.
[b] The seed of the fruit of an East Indian tree used as a source of strychnine.

animal. Doctor Findlay Russell, who has made enormous contributions to our understanding of the nature of animal toxins, their modes of biological action, and the procedures for treating people who have been envenomed or poisoned, estimates that there are about 1200 known species of poisonous or venomous marine animals, "countless" numbers of venomous arthropods (spiders), and about 375 species of dangerous snakes (out of a total of about 3500 species).

We have been speaking of both "venomous" and "poisonous" animals, and there is a distinction between the two. A venomous animal is one that, like a snake, has a mechanism for delivering its toxins to a victim, usually during biting or stinging. A poisonous animal is one that contains toxins in its tissues, but cannot deliver them; the victim is poisoned by ingesting the toxin-containing tissue.

An interesting and important example of an animal poison is paralytic shellfish poison (PSP). This chemical, which is also known as saxitoxin and by several other names as well, is found in certain shellfish. But it is not produced by shellfish; it is rather a metabolic product of certain marine microorganisms (Protista). These microorganisms are ingested by the shellfish as food, and their poison can remain behind in the shellfish's tissue. Paralytic shellfish poison is not a protein, but a highly complex organic chemical of most unusual molecular structure.

Shellfish accumulate dangerous levels of PSP only under certain conditions. Typically, this occurs when the microorganisms undergo periods of very rapid growth, resulting from the simultaneous occurrence of several favorable environmental conditions. This growth, or "bloom," frequently imparts a red color to the affected area of the ocean, and is referred to as a *red tide*. Shellfish growing in a red tide area can accumulate lethal amounts of PSP.

Red tides (and some with other colors as well) occur with some regularity in certain coastal waters of New England, Alaska, California, and several other areas. If it is the type of tide that can produce PSP or other toxins, public health officials typically quarantine affected areas to prevent harvesting of shellfish. In some areas of the Gulf of Alaska, large reservoirs of shellfish cannot be used as food because of a persistent PSP problem.

Paralytic shellfish poison, like botulinum toxin, is a neurotoxic substance and can also affect certain muscles, including the heart, which are under nervous system control. Some poisoned humans who have recovered from the effects of PSP have described the early stages of intoxication as not at all unpleasant: a tingling sensation in the lips and face and a feeling of calm. Those who die from PSP ingestion do so because of respiratory failure.

The first successful method for measuring the amount of PSP in shellfish was published in 1932. The procedure used was a bioassay. The bioassay for PSP was simple. Extracts from shellfish suspected of contamination were fed to mice. The poison was measured in "mouse units." A mouse unit was the amount of toxin that would kill a 20 gram mouse in 15 minutes. Crude, but nevertheless effective at telling public health officials when shellfish were too toxic to eat.

The plant kingdom is another source of some unusually toxic chemicals. A few examples are presented in Table 4.3, along with a description of some of their biological effects.

Infants and preschoolers are the most frequent victims of plant toxins. Their natural curiosity leads them to put all sorts of non-food items into their mouths, and berries, flowers, and leaves from house and yard plants are often attractive alternatives to spinach. The number of deaths from consumption of poisonous plants is not great, but the number of near-deaths is; about 10% of inquiries to poison control centers concern ingestion of house, yard, and wild plants, including mushrooms. Among the house plants dumbcane (species of dieffenbachia) and philodendrons are prominent, and a fair number

Table 4.3 *Poisonous properties of some common plants*
(The specific chemicals involved are in some cases not known)

| Plant | Effects |
|---|---|
| Water hemlock | Convulsions |
| Jimson weed | Many, including delirium, blurred vision, dry mouth, elevated body temperature |
| Foxglove, lily-of-the-valley, oleander | Digitalis poisoning – cardiovascular disturbances |
| Dumbcane (dieffenbachia) | Irritation of oral cavity |
| Jonquil, daffodil | Vomiting |
| Pokeweed | Gastritis, vomiting, diarrhea |
| Castor bean | Diarrhea, loss of intestinal function, death |
| Poison ivy, poison oak | Delayed contact sensitivity (allergic dermatitis) |
| Potatoes, other solanaceous plants[a] | Gastric distress, headache, nausea, vomiting, diarrhea |

[a] The Solanaceae include many species of wild and cultivated plants, the latter including potatoes, tomatoes, and eggplants. All these plants contain certain natural toxicants called solanine alkaloids. The levels found in the varieties used for food are below the toxic level, although not always greatly so. Potatoes exposed to too much light can begin to grow and to produce excessive amounts; the development of green coloring in such potatoes (chlorophyll) indicates this growth. Storing potatoes under light, particularly fluorescent light, is to be discouraged.

of poisonings arise from jade, wandering Jew, poinsettia, schifflera, honeysuckle, and holly.

Children are also especially vulnerable for a reason touched upon in Chapter 2. Consider the family of mushroom toxins known as amatoxins. Almost all mushroom-related deaths in North America are caused by these toxins, which are metabolic products of *Amanita phalloides*. These toxins are slightly unusual because symptoms appear only after 12 hours following ingestion; they include vomiting, diarrhea, and very intense abdominal pain. Ultimately the toxins cause liver injury that can be serious enough to cause death.

The lethal dose of amatoxins is in the range of 1 mg/kg b.w. For an adult weighing 70 kg, a total of about 70 mg needs to be consumed to cause death (1 mg/kg × 70 kg). For a one-year-old child weighing

10 kg, only about 10 mg needs to be ingested to create a life-threatening intake of the toxins. Small body size is highly disadvantageous.

## Synthetic poisons

Perhaps the most prominent and well-studied class of synthetic poisons are so-called *cholinesterase inhibitors*. Cholinesterases are important enzymes that act on compounds involved in nerve impulse transmission – the neurotransmitters (see the later section on neurotoxicity for more details). A compound called acetylcholine is one such neurotransmitter, and its concentration at certain junctions in the nervous system, and between the nervous system and the muscles, is controlled by the enzyme acetylcholinesterase; the enzyme causes its conversion, by hydrolysis, to inactive products. Any chemical that can interact with acetylcholinesterase and inhibit its enzymatic activity can cause the level of acetylcholine at these critical junctions to increase, and lead to excessive neurological stimulation at these *cholinergic junctions*. Typical early symptoms of cholinergic poisoning are bradycardia (slowing of heart rate), diarrhea, excessive urination, lacrimation, and salivation (all symptoms of an effect on the parasympathetic nervous system). When overstimulation occurs at the so-called neuromuscular junctions the results are tremors and, at sufficiently high doses, paralysis and death.

Two important classes of cholinesterase inhibitors are the organophosphates and the carbamates, a few of which are widely used insecticides. Two such insecticides are chloropyrifos and carbaryl (structures shown). They are highly effective insecticides and, if used properly, appear to be without significant risk to humans (although the use of chloropyrifos and some other members of the class is somewhat controversial).

Chloropyrifos

Carbaryl

Interestingly, much work has been devoted to the development of substances in the class of cholinesterase inhibitors that have exceedingly high toxicity; substances that also have properties (such as volatility and sufficient but not excessive environmental stability) that make them useful as agents of warfare. Most of those now stockpiled were first developed during World War II. Sarin and VX are perhaps the most well-known members of this class of compounds that have been especially designed to kill people.

Sarin
(nerve gas)

VX
(nerve gas)

Of course these and other "nerve gas" agents are available to terrorists. Sarin, for example, was used twice during the 1990s by Japanese terrorists in Matsumato and Tokyo. Not only were the typical symptoms of acute poisoning, including death, observed, but follow-up studies revealed a number of serious delayed effects. The question of delayed effects remains significant for the whole large class of cholinesterase inhibitors. Drugs such as atropine are used to treat anticholinesterase poisoning; atropine reduces the effects of excessive acetylcholine by altering the particular receptors at which it normally acts to transmit nerve impulses.

As in the case of botulinum toxins, certain cholinesterase inhibitors are effective medicines. Physostigmine, a naturally occurring carbamate derived from the calabar bean, has been used as a glaucoma treatment since the late nineteenth century. Neostigmine, another carbamate, is used to overcome the acetylcholine deficiency that is the cause of *myasthenia gravis*. Another anticholinergic agent called tacrine is effective at alleviating dementia in some subtypes of Alzheimer's disease. All of the medicines can, at high doses, cause poisoning. Again, old Paracelsus is seen to be correct.

## Poisonings from industrial accidents

Although industrial chemicals, or the incidental by-products of industrial society, do not create quite the risk of acute lethality that some natural and deliberately synthesized poisons do, there are some that can cause considerable toxicity after a single exposure. Most of these acute exposures are created by industrial or transportation accidents, which are not infrequent. To minimize the damage accidental releases such as these might cause, it has become important to use the toxicity rating classification discussed earlier and to label industrial chemicals accordingly to ensure appropriate care is taken in handling, storing, and transporting them. Of course, there have been industrial accidents involving releases sufficiently large to cause death, sometimes to workers, sometimes to nearby residents or passers-by. The worst example of this type of event took place during the night of December 3, 1984, at Bhopal, India. Approximately 40 tons of methyl isocyanate (a very simple organic chemical used in the synthesis of an important pesticide and having the structure shown) were released into the atmosphere, killing more than 2000 people and injuring many more.

Methyl isocyanate

For a series of chemicals that are commonly involved in accidental release of one type or another, the EPA has developed Acute Exposure Guidelines, called AEGLs. They are intended for use in guiding actions under emergency conditions, and are expressed as air concentrations that are associated with certain adverse outcomes after specific and relatively short-term exposure periods, from minutes up to 8 hours. The agency has found three types of AEGLs useful: one that describes the air concentration producing effects that are non-disabling (AEGL-1); those that specify the air concentration that is disabling and which could prevent people from removing themselves from the affected area (AEGL-2); and the concentration that could be life-threatening (AEGL-3). The AEGLs are derived from available information obtained from the study of accidental and occupational exposures, and from animal experiments.

Although there have been a number of industrial releases of hydrogen chloride (HCl) that have led to life-threatening effects or death, it has not been possible to collect reliable information on the concentration – time relationships for these outcomes. So the EPA has relied upon animal data to derive an AEGL-3 for this important compound. It was estimated that for periods of one hour, levels of 160 mg/m$^3$ could cause life-threatening effects, and this became the AEGL-3 (one hour). Something called Haber's Law seems to hold for very short term exposures to airborne poisons. The Law (perhaps it is more like a reasonably well-documented rule of thumb) says that for short-term exposures, not to exceed 8 hours, $C \times T = $ a constant. Here $C$ is the airborne concentration of the agent and $T$ is the duration of exposure in hours. That is, an AEGL-3 of 160 mg/m$^3$ for one hour ($C \times T = 160$) is equivalent to an AEGL-3 of 320 mg/m$^3$ for one-half hour, or an AEGL-3 of 40 mg/m$^3$ for four hours. This rule is used frequently because the data available on acute exposure outcomes is usually insufficient to specify the AEGL's for as many different time periods as we would like. So, in most cases, the available data are used to derive an AEGL for one time period, and the Haber Law is used to derive AEGLs for all the time periods that are thought useful.

A non-disabling effect is typically some type of mild irritation to the eyes or upper respiratory tract. For another common cause of accidental poisoning, ammonia ($NH_3$), the AEGL-1 for 30 minutes is 17 mg/m$^3$. The AEGL-2 for ammonia for this same time period has been set at 112 mg/m$^3$, and it is a level that can cause severe eye and throat irritation; although the effects are still reversible at this level they are severe enough to prevent people from acting as they normally would to escape danger. The AEGL-3 for ammonia has been set at 1120 mg/m$^3$ for 30 minutes.

The AEGLs are intended for use in guiding emergency actions when members of the general population are involved. Other guidelines are available that are intended for use in occupational settings. These include the so-called Short-Term Exposure Limits (STELs) developed by the American Conference of Governmental Industrial Hygienists (ACGIH), and the levels Immediately Dangerous to Life and Health (IDLH) put forth by the National Institute of Occupational Safety and Health. These occupational limits are not enforceable regulatory standards, but are widely used. As we shall see in the chapters on risk assessment, there are often differences between health protection standards meant to apply to the general population, and those applicable to workers, because workplace populations are generally not considered to include people with extreme susceptibility. These differences can be seen in some short-term exposure limits as well as those used to protect against other forms of toxicity.

There are hundreds of extraordinarily interesting tales about highly toxic chemicals, but to say more on this topic gets us too far adrift from our main course. A couple of important points have been made here. First, virtually all chemicals can cause a deadly response after a single dose, but enormous differences exist among chemicals in their capacity to kill. The dose–response relationship for acute lethality, from which we derive $LD_{50}$ or other measures of lethality, is one common measure of acute toxicity. As we learned in Chapter 3, it is common to observe large species differences in $LD_{50}$ values for the same chemical; variability in response among species is the general rule for both acute and other forms of toxicity. Categorizations of chemicals as "extremely toxic", "practically non-toxic", and so on, generally refer only to a very limited aspect of their toxic potential, and should not be used as general indicators of the full range of toxic effects a chemical may cause.

We shall now move to a more complete picture of chemical toxicity than can be gained from a look at the acute poisons.

## Slow poisons

The rest of this chapter deals with "slow poisons," but this title, while conveying a message that has a popular meaning, is a little misleading. A more accurate title might be "Toxicity associated with doses that do not give rise to immediately observable adverse effects." This statement avoids the false impression conveyed by a title that suggests there are two categories of chemicals, those that are acutely toxic and those that are "slow poisons." From the principles discussed so far, it should be obvious that all chemicals can be both "fast" and "slow" poisons, depending upon the size, duration, and other conditions of dosing. "Slow poisoning" is perhaps a better title.

Slow poisoning can occur in several different ways. In some cases, chemicals or their metabolites may slowly accumulate in the body – rates of excretion are lower than rates of absorption – until tissue and blood concentrations become sufficiently high to cause injury. Delayed toxicity can also be brought about by chemicals that do not accumulate in the body, but which act by causing some small amount of damage with each visit. Eventually, these small events, which usually involve some chemical interaction between the visiting chemical and normal cellular constituents, add up to some form of injury to the organism that can be observed and measured by the toxicologist.

Another possible mechanism of delayed toxicity involves creation of some serious form of cellular damage, involving the cell's genetic machinery, as a result of one or a very few chemical exposures. The damage may be passed on within the cell's genetic apparatus, to future generations of cells, even if the chemical causing the initial damage never again appears in the body. The reproducing, but deranged cells, if they survive, may eventually create a disease state, such as cancer, in the host.

Knowledge of which mechanism of delayed toxicity is operating in specific cases cannot usually be gained from the animal test or from epidemiology studies; additional studies of ADME, and of pharmacodynamic interactions of the chemical with cellular components, are necessary to understand mechanisms of delayed toxicity. Some mechanisms are discussed in the following to illustrate the value of this kind of study.

## Some slow poisons and their targets

The toxic properties of the chemicals to be discussed have been learned from the many types of general and specialized animal tests discussed in Chapter 3. In many cases they have also been learned from epidemiological studies and case reports. Carcinogens, as we have already mentioned, are excluded until the next two chapters.

There are several possible ways to categorize chemical toxicity. Perhaps the most common is by grouping chemicals according to the targets they can damage, and it is the approach followed here.

One result of this approach is that, from the chemist's point-of-view, the grouping is highly heterogeneous. Thus, under liver toxicants are grouped a number of organic solvents, some metals, certain pesticides, some naturally occurring chemicals, a few pharmaceutical agents, and a miscellaneous collection of industrial chemicals of diverse structural properties. In fact, each target group will contain such an assortment. What is seen when toxicity is grouped by target is that substances of diverse structural, chemical, and physical properties can affect the same biological target. And although chemicals having highly similar structures tend to produce similar forms of toxicity, there are many dramatic examples in which small modifications in chemical structure can result in substantial shifts in toxic potency and even toxicity targets.

In a few cases, the way a chemist might group chemicals does match, to a degree, the way the toxicologist groups them. Chemicals that can dissolve fatty materials, and that are used as solvents for them, tend to have similar physical properties. Many of these chemicals can impair the nervous system by a biological mechanism that depends upon their characteristics as solvents. But this type of matching is not the general rule.

The three targets that are the first point of contact between environmental chemicals and the body will be discussed first: the *gastrointestinal tract*, the *respiratory system*, and the *skin*. Recall from Chapter 2 that chemicals enter the *blood* after absorption, so this fluid is the next target (see Figure 2.1). Then come the *liver*, the *kidneys*, and the *nervous system*. The chapter concludes with a discussion of some chemicals that can damage the *reproductive system* and some that can cause birth defects, the so-called *teratogens*, and other forms of *developmental toxicity*. Brief discussions of *immune system*, *cardiovascular system*, *muscle*, and *endocrine system toxicities* are also offered.

Only a few, well-known examples of chemicals that can damage these targets are presented; for most of the targets a complete list would include several dozen up to several hundred substances! Notice also that some chemicals appear on two or three lists. The chemicals reviewed were selected primarily because they provide good illustrations of various toxicological phenomena, and not necessarily because they are environmentally important (although some certainly are).

The critical question of dose–response relationships is given only cursory mention in this chapter. Keep in mind that all of the toxic phenomena described in this chapter and those on carcinogens exhibit such relationships; we return to the dose–response issue in the chapters on risk assessment.

Also, keep in mind that because a chemical is listed as producing toxicity does not mean it produces this toxicity under all conditions of exposure. Whether a liver toxicant is likely to produce its effects in human beings under their actual conditions of exposure is only partially answered by the knowledge that it has been shown to cause liver toxicity in test systems, or in certain groups of highly exposed people. A full risk assessment is needed to answer the ultimate question, and a great deal more must be known before that question can be dealt with.

A final note of caution: the toxicity information provided on individual chemicals is by no means complete. Data have been selected to illustrate certain principles; no attempt is made to provide anything close to a thorough toxicological evaluation of any of the chemicals discussed. The *Sources and recommended reading* section, appearing after the final chapter, lists several authoritative sources of toxicity information on individual chemicals.

## Respiratory system

The various passages by which air enters the body, together with the lungs, comprise the respiratory system. Its principal purpose is to move oxygen into the blood, and to allow the metabolic waste product, carbon dioxide, to exit the blood and leave the body in exhaled air. The exchange of these two gases occurs in the lung. The respiratory system serves other purposes as well, and includes a mechanism for the excretion of toxic chemicals and their metabolic products.

The so-called respiratory tract has three main regions. The nasopharyngeal region includes the nasal passages and the pharynx, which is a cavity at the back of the mouth; these passages are the first point of contact for air and chemicals carried within it, in the form of particles, gases, or vapors. Below the pharynx lies the second, tracheobronchial region, which includes the trachea (windpipe), at the top of which sits the larynx (voice box); off the windpipe extend two tubes called bronchi, one leading to each lung. The bronchi undergo several branchings within the lung and finally lead to the *pulmonary* region. Here the branches lead to bunches of tiny air sacs, which in turn end in small "pouches" called alveoli, where oxygen entering the body and carbon dioxide exiting change places. There are about 500 million alveoli in the adult human being, with a total surface area of about 500 square feet! The two bronchi and their many branches can be thought of as inverted trees extending into the lungs, and the air sacs ending as alveoli, as leaves on those trees. Although the air within these respiratory structures is ostensibly "inside" the body, in fact it is not; all of it is connected without obstruction to the air outside the body, and components of air, including oxygen, get truly inside the body – into the bloodstream – only after they pass through the cells lining the alveoli.

Gases, vapors, and dust particles can move into the airways from the environment and penetrate to the pulmonary region in several ways. Gases and vapors can move readily through the three regions, but dust particles are blocked at several points along the way. Large particles, those typically sneezed out, do not get beyond the nasopharynx. The trachea and bronchi are lined with epithelial cells (*epithelial* is the adjective biologists attach to cells that act as linings within the body), that secrete mucus and that also hold little hair-like attachments called cilia. The cilia and mucus can collect particles that are small enough to negotiate their way through the nasopharyngeal region, and the cilia move those particles up to the mouth, where they collect in saliva, either to be excreted or swallowed. Some very tiny particles, generally those less than one micrometer (μm) (one-millionth of a meter) in diameter, instead of being caught in the "mucociliary escalator," as it is called, manage to get eaten up or engulfed by cells called phagocytes. Phagocytes carry dust particles into various lymph nodes, whereby they can enter the blood. Gases and vapors can, to varying degrees depending upon their chemical and physical properties, be absorbed into the blood at any of the three regions of the respiratory tract, but most absorption takes place in the pulmonary region. Some of these substances cause systemic toxicity, others only cause local toxicity in the respiratory tract; and others can cause both types of toxicity.

Toxicologists who study the responses of the respiratory system to foreign chemicals generally categorize those responses according to their biological and pathological characteristics.

*Irritation* is caused by many chemicals, including the common gases ammonia, chlorine, formaldehyde, sulfur dioxide, and dust containing certain metals such as chromium. The typical response to a sufficiently high level of such substances is constriction, or tightening, of the bronchi; this is accompanied by *dyspnea* – the feeling of being incapable of catching the breath. With the airways constricted in this fashion, oxygen cannot get to the pulmonary region at a sufficient rate to satisfy the body's demands. This type of constriction brings on asthma attacks and, if chronic, a long-lasting bronchitis, or inflammation of the bronchi, may ensue. Sometimes a serious swelling, or *edema*, occurs in the airways, and when irritation is particularly serious, it can pave the way for microorganisms to invade the tissue of the airways and lungs, to cause an infection. A lethal exposure is one that completely overwhelms the responsive power of the respiratory system, and turns off the respiratory process for good. Generally, however, irritation and edema subside following cessation of exposure – they are often reversible phenomena.

Exposure to irritating levels of gases such as ammonia, chlorine, and hydrogen chloride, and metals such as chromium, usually occur only in certain occupational settings, although occasionally the general population becomes exposed because of a transportation or industrial mishap (although solutions of ammonia are sold as household products, and almost everyone has experienced in a mild way the pungent qualities of ammonia gas). Sulfur dioxide is a somewhat different matter. This gas, $SO_2$, is produced by the burning of fossil fuels, all of which contain some sulfur. Metal smelting operations also produce $SO_2$ as a by-product, because of the presence of sulfur in the raw ores. These burning and smelting processes also produce particles containing $SO_2$ reaction products, called sulfates. As a result, sulfur dioxide and fine particles containing various sulfates are common air pollutants to which millions are exposed on a daily basis. At some times and in some places levels can shoot up and cause disturbingly unpleasant irritating effects, as well as bronchoconstriction. An especially serious pollution event is the so-called "sulfurous smog" caused by the accumulation of dense particles containing sulfuric acid, $H_2SO_4$. The most serious of such smogs occurred in London in 1952, where around 4000 people died because of the event. Twenty deaths occurred in 1948 in Donora, Pennsylvania, because of sulfurous fog.

People who suffer from other pulmonary diseases that interrupt the flow of oxygen are especially sensitive to the irritating effects of $SO_2$ and its particulate derivatives. This gas and several other gaseous air pollutants, to be mentioned in a moment, can cause other, delayed toxic effects in the respiratory system. Note also that these same chemicals are the principal causes of acid rain.

Some especially irritating organic chemicals are certain gaseous or highly volatile compounds called aldehydes, the most well-known of which is formaldehyde. This gas is a natural product of combustion, and is present in smoke, including that from tobacco. This and a few related aldehydes are the principal agents causing irritation in the upper respiratory region when smoke is inhaled. Formaldehyde is, of course, a major industrial chemical (7 000 000 tons produced in 1999) that is used to manufacture plastics, including an insulating material, urea–formaldehyde foam, that has been installed in millions of homes; residual levels of the gas may emanate from these materials and be irritating to some individuals. The chemical is also used to manufacture "sizing" for synthetic fabrics, a process that gives a permanent press to certain clothing. Because of its natural occurrence and industrial production formaldehyde is omnipresent in both indoor and outdoor atmospheres.

$$\begin{array}{c} H \\ \diagdown \\ C{=}O \\ \diagup \\ H \end{array}$$

Formaldehyde

A second category of respiratory toxicity is that characterized by damage to the cells anywhere along the respiratory tract. Such damage can cause the release of fluid to the open spaces of the tract, and result in accumulation of that fluid, or edema, in several areas. These edematous reactions can occur after acute exposure to some chemicals, although the production of edema can be delayed, and arise after subchronic and chronic exposures.

Two common and widespread air pollutants, ozone ($O_3$), a potent oxidizing agent, and nitrogen dioxide ($NO_2$), are good examples of chemicals that can cause cellular damage in the airways and lungs. Sustained exposure to these gases can cause *emphysema*, with accompanying loss of capacity for respiratory gas exchange; this condition can, of course, lead to serious physical disability. The effects of these gases are compounded by smoking.

The Clean Air Act recognizes a number of so-called "primary air pollutants," and the EPA has established standards for these substances. Ozone, nitrogen oxides, and sulfur dioxide are among these (the others are carbon monoxide and lead, discussed below, and "total suspended particulates"). The EPA's standard for ozone is 0.08 parts of the gas per million parts of air (0.08 ppm), averaged over eight hours. Standards also exist for the oxides of sulfur and nitrogen. These are designed to prevent chronic respiratory toxicity of any kind.

Air quality is significantly impaired by the presence of particulate matter (PM), and the importance to public health of these small particles of varying chemical composition has greatly increased in the past decade. Particulate matter has been transformed, through a series of complex epidemiological studies, from a mere "nuisance" to a cause of serious pulmonary and cardiovascular disease, particularly for highly susceptible individuals. The composition of PM varies geographically depending upon local sources, but it is generally produced by combustion, and by both industrial and natural processes that generate dusts of widely ranging chemical composition. The mechanisms whereby

PM causes its health effects, especially its effect on cardiovascular health, are unclear; there is some evidence that chemical composition of PM is not as important as particle size, with the greatest risks associated with what is designated as PM 2.5 (particle size less than 2.5 μm). Some experimental evidence suggests (but does not establish) that so-called fine (0.25 to 1.0 μm) and ultrafine (<0.25 μm) are the most potent toxicants, but regulation is now focused on PM 2.5. Regulation of PM and the primary air pollutants is highly contentious, because the costs of controlling them are enormous.

Some occupational situations, if inadequately controlled, can create opportunities for damage to the respiratory system. Exposure to certain forms of the metals nickel and cadmium, ordinarily as airborne particulates, can cause cellular damage, edema, and, if sustained for sufficiently long periods, emphysema. Many other metals, usually only in some of their many chemical forms, can produce emphysema upon subchronic or chronic exposure.

A particularly interesting example of a respiratory toxicant is the pesticide called paraquat. This organic chemical produces serious and generally delayed pulmonary edema after *ingestion*. This example illustrates the phenomenon of systemic toxicity – toxic effects at sites of the body distant from the site of initial contact and which can be reached only after a chemical enters the bloodstream – and further that the lungs can be not only a site of direct, local toxicity, but also a target for systemic effects. Paraquat, after being absorbed through the gastrointestinal tract, enters the circulatory system and thereby reaches an organ where it is particularly active, the lungs.

*Fibrosis* is a third category of pulmonary damage. Certain particles and dusts, when inhaled for long periods in sufficiently fine particle size, can create cellular damage in the lungs of a type that causes those cells to exude fibrous materials, much like tiny filaments of connective tissue or gristle. If there is a sufficient build-up of these fibers, the lung tissue can become rigid and lose function. In its advanced form fibrosis is a serious and debilitating disease.

Fibrosis was first recognized in certain occupational settings. One of the well-known conditions of this type is silicosis, which is brought about by long-term, uncontrolled exposure to certain crystalline forms of silica ($SiO_2$), and certain related substances called silicates. These minerals are widespread on earth, in fact most of the inorganic, non-aqueous earth consists of silica and silicates. Many of these minerals (e.g., quartz) have major industrial uses. It is important to emphasize that silica and silicates occur in both crystalline and non-crystalline (amorphous) forms, and it is only the former that causes silicosis. Occupational exposures to crystalline silica and silicates have to be carefully controlled.

Asbestos is the name given to several different fibrous forms of silicates; these go under names such as crocidolite, amosite, and chrysotile. These forms are, to varying degrees, also capable of eliciting fibrosis in lung tissue. In Chapter 6 it will be seen that some forms are capable of more serious damage – cancers of the lung and the mesothelium.

The final category of respiratory response worth noting includes some allergies. Allergic reactions are a special brand of adverse effect, resulting from the *immune system's* response to many types of foreign agent, including microorganisms, certain large protein molecules, and even some relatively small foreign molecules. The first exposure to an antigenic chemical may result in the interaction of that chemical with certain normal proteins to form complexes called antigens; antigens in turn provoke the immune system to form other complex entities called antibodies, which remain in the body. No allergic reaction, but only those biochemical changes take place as a consequence of this first exposure.

But subsequent exposure to the same chemical can be for some people quite devastating. This exposure results in an interaction between newly formed antigens and the antibodies that were produced from the first exposure, which in turn elicits a series of biochemical and physiologic responses ranging from mild flushing of the skin, all the way to death. One chemical that produces allergic-like responses in the lungs is toluene diisocyanate (TDI), a volatile chemical used to produce polyurethane plastics. It is a major industrial chemical, and worker exposure needs to be tightly controlled. The mildest manifestation of a TDI-induced allergenicity is bronchoconstriction, but at sufficiently high exposures some sensitive individuals can suffer substantial losses in pulmonary function from which they only slowly recover.

The allergic response differs from the usual toxic response in that a prior exposure is necessary to create the conditions for an adverse response.

### Gastrointestinal tract and skin

The two other body areas that are first contact points for environmental chemicals are the gastrointestinal (GI) tract and the skin. The GI tract is that long and many-faceted tube beginning at the mouth and extending downward as the pharynx and the esophagus, then enlarging as the stomach, narrowing again as the small intestine, and ending as the large intestine (which consists of the cecum, colon, rectum, rectal canal and anus). The GI tract is, as shown in Chapter 2, a major point of entry for environmental chemicals present in food and water, and even in soils and dusts. It is relatively rare that environmental chemicals reach sufficiently high levels in these media to produce significant toxicity directly in the GI tract, although accidental ingestion of many chemicals may cause severe injury to it. Highly caustic materials such as lye (sodium hydroxide) have been accidentally ingested by many individuals and have been shown capable of causing serious damage to the lining of the GI tract. Such materials, because of their strongly alkaline properties, essentially destroy the natural fatty chemicals present in the cell membranes, either of the GI tract, or at any other site of direct contact. But doses below those causing such readily noticeable effects appear not to cause any adverse reactions upon subchronic or chronic exposure.

While most cases of GI tract distress are probably due to microbiological agents or their toxins, many chemicals are capable of inducing vomiting and diarrhea and other GI tract responses. Poisoning with heavy metals such as lead, cadmium, and arsenic may be suspected in patients reporting with severe abdominal pain, nausea, and vomiting, although these are typically the consequence of acute, high-level exposures. In many cases the GI tract response leading to vomiting or diarrhea is an indirect effect of the chemical, secondary to a systemic attack on the nervous system which controls GI tract behavior. In contrast, certain microbiological agents, such as species of *Salmonella* present in contaminated food or water, may grow in the GI tract and directly induce these effects.

Perhaps the most important toxicological role played by the GI tract is its influence over the absorption of chemicals that enter it. That absorption rates vary widely among chemicals has been explained in Chapter 2, but how the GI tract and its contents contribute to this phenomenon was not explained. Mechanisms of absorption are many and varied, and are influenced by the type and quantity of food present at the time of chemical ingestion, the pH (degree of acidity) of various portions of the GI tract, and even the nature and activity of the microorganisms that normally live in the intestines. In fact, metabolism of certain chemicals brought about by these microorganisms can play a crucial role, not only in their absorption, but also in the nature of the systemic toxicity they ultimately produce.

*Skin* is our largest organ. The average adult's body surface area is about nine square meters and the skin of which it is comprised weighs 20–30 pounds (about 15% of body weight)! All sorts of chemical agents come into direct contact with the skin. A few are metabolized in parts of it, and many pass through it into the circulation.

Because adverse skin responses are so easily recognizable, this organ was among the earliest subjected to scrutiny, mostly by physicians interested in occupational diseases. Bernardino Ramazzini's tract of 1700, *De Moribis Artificum Diatriba*, contained many examples of skin diseases associated with occupational exposures, and, as will be seen in the next chapter, the seminal work of Percival Pott on occupationally induced cancers, published in 1775, revealed the role of soot in the production of cancers on the skin of the scrotum in London chimney sweeps.

Irritation of the skin is brought on by a very large number of chemicals. It is characterized by reddening, swelling, and itching, which generally subside after exposure ceases. Allergic responses, as in the case of those occurring in the respiratory system, require prior sensitization; subsequent exposures bring on an attack. Formaldehyde causes allergic responses in the skin of sensitive individuals at exposures lower than those necessary to elicit irritation.

No doubt the greatest environmental threat to the skin is not chemical, but is rather a physical agent, sunlight; most skin cancers are caused by excessive exposure to ultraviolet radiation from the sun.

### Blood and lymphatics – hematoxicity and immunotoxicity

Nutrients absorbed from the gastrointestinal tract and oxygen from the lungs enter the blood stream and are thereby carried throughout the body, where they feed the machinery of cells. Blood is, of course, pumped by the heart through the *circulatory system*, a complex network of tubes, called vessels. There are three main types of blood vessel: the *arteries* carry blood containing nutrients and oxygen away from the heart and the *veins* carry blood and cellular waste products including carbon dioxide back to it; there is also present a system of small, thin-walled tubes called *capillaries* which branch off the arteries and subsequently merge to form veins.

Blood *plasma* is the liquid, mostly water, that carries several types of blood cells as well as nutrients, and other chemicals, such as hormones, that need to be transported around the body.

In Chapter 2 we explained how chemicals foreign to the body can enter the circulatory system and be transported to various parts of the body. If the concentrations of these substances or their metabolic products reach sufficiently high levels, systemic toxicity can result. Different chemicals affect different organs and systems of the body because of differences in the rate and manner of their absorption, distribution, metabolism, and excretion. Some chemicals are directly toxic to the elements of the blood. Others bring about changes in certain elements of the blood that become detrimental to other systems of the body.

The most well-known example of the latter is carbon monoxide. This simple gas (CO) is a product of incomplete combustion of organic substances. Complete combustion results in the conversion of carbon-containing chemicals, such as the hydrocarbons used as fuels, to carbon dioxide ($CO_2$), but because most combustion systems cannot allow for the presence of all of the oxygen needed, some of the carbon ends up in the less oxidized form of CO. Incomplete combustion of gasoline in trucks and automobiles is one of many sources of this gas. Individuals who are unable to escape from fires can be exposed to very high levels of CO, and many fire deaths are related to this gas. Firefighters, garage workers and traffic policemen can experience relatively high concentrations while on the job, and the rest of us inhale the gas throughout the day because of its ubiquitous presence in the atmosphere. Smokers get an additional dose. Carbon monoxide gradually oxidizes to $CO_2$ and so does not continue to accumulate in the atmosphere.

The gas has a molecular size and shape similar to oxygen ($O_2$). When oxygen passes from the lungs into the blood it interacts with a large molecule called *hemoglobin* (Hb). This vital chemical is present in the red blood cells (*erythrocytes*). In addition to a large protein component Hb contains a complex organic compound called heme; the heme molecule carries within it an ion of the inorganic element iron. Under normal circumstances oxygen molecules, after they pass through the lungs, interact with Hb, specifically with the iron-heme portion.

$$Hb + O_2 \rightarrow O_2Hb$$

The $O_2Hb$ molecule is called oxyhemoglobin and has a bright red color. The red blood cells transport $O_2Hb$ to all cells of the body, where $O_2Hb$ dissociates, yielding up the needed $O_2$ molecules.

Carbon monoxide is dangerous because, like oxygen, it has an affinity for Hb, and produces a bright, cherry-red substance called carboxyhemoglobin.

$$Hb + CO \rightarrow Carboxyhemoglobin$$

In fact, CO's affinity for Hb is even greater than that of oxygen, by several hundred times! Because the body's supply of red blood cells and Hb is limited, the presence of CO in inhaled air can deprive the body of oxygen, a condition called *anoxia*. The nature, duration, and severity of the resulting toxicity depend upon the blood COHb level created which, of course, depends upon the concentration of the CO in the inhaled air and the length of time the air is inhaled. The presence of COHb in the capillary blood imparts an abnormal red color to skin and fingernails. The conditions creating toxicity arise in the blood; the actual effects appear in the nervous system, in the heart, and elsewhere.

One of the most carefully worked out dose–response relationships is that for carbon monoxide poisoning. Based on controlled studies of exposure in humans at low levels and on observations in humans who have suffered high level exposures because of their occupation or because of accidents or suicide attempts, the relationship between blood levels of carboxyhemoglobin (COHb) and toxicity is understood as follows:

| Per cent COHb in blood | Signs and symptoms |
| --- | --- |
| 0–10 | *See text, below |
| 10–20 | Headache |
| 20–30 | Headache, throbbing in temples |
| 30–40 | Headache, dizziness, nausea, vomiting, dimness of vision |
| 40–50 | Collapse, increased pulse rate |
| 50–60 | Increased respiration, coma |
| 60–70 | Coma, convulsions, depressed heart rate |
| 70–80 | Respiration severely depressed, death within hours |
| 80–90 | Death within one hour |
| 90–100 | Almost immediate death |

Except for death and possibly damage to the heart, these effects are reversible when the CO source is removed, because COHb eventually dissociates and releases CO to the lungs, where it can be excreted. The treatment for CO poisoning is administration of oxygen to hasten dissociation of the COHb molecule.

No human health effects have been detected at COHb blood levels below about 2% ("background" levels in non-smokers average about 0.5%). Subtle effects on the nervous system, such as reduced ability to sense certain time intervals, have been reported at blood levels of 2.5%. At COHb levels of 5% certain cardiovascular changes are detectable, especially in patients with coronary heart disease. Heavy smokers exhibit COHb levels in the range of 5–6%, and if they happen to be pregnant, the fetus can suffer the effects of oxygen deprivation.

The quantitative relations between blood levels of COHb and air levels of CO have been well worked out. In general, the blood level achieved is a function of both air concentration and the length of time the individual breathes the air ($C \times T$). Legal limits on workplace and environmental concentrations have been established to avoid significant COHb levels in the blood, but of course this goal is not always realized.

The hemoglobin molecule can be adversely altered in other ways by certain chemicals. The ion of iron that is present can be oxidized by certain chemicals to produce a brown-black compound called *methemoglobin*. The latter cannot bind to oxygen, so chemicals creating it can produce serious toxicity if the concentrations generated are sufficiently high. Nitrites, inorganic ions having the structure $NO_2^{-1}$, are particularly successful at creating methemoglobinemia (excess methemoglobin in the blood).

In addition to erythrocytes, blood contains white blood cells, called leukocytes, of several types, and platelets, also called *thrombocytes*, which control blood clotting. Hematopoiesis (from the Greek, "haimo," for blood, and "poiein" for "to make") is the process by which the elements of the blood are formed. The marrow of bone contains so-called stem cells which are immature predecessors of these three types of blood cells. Chemicals that are toxic to bone marrow can lead to *anemia* (decreased levels of erythrocytes), *leukopenia* (decreased numbers of leukocytes), or thrombocytopenia. *Pancytopenia*, a severe form of poisoning, refers to the reduction in circulatory levels of all three elements of the blood. One or more of these conditions can result from sufficiently intense exposure to chemicals such as benzene, arsenic, the explosive trinitrotoluene (TNT), gold, certain drugs, and ionizing radiation. Health consequences can range from the relatively mild and reversible to the severe and deadly. Some chemicals produce an excess of certain of the blood's elements and this may signal equally serious consequences for the health of the affected person. Such is the case with the various *leukemias*, characterized by greatly increased numbers of certain leukocytes, that have become abnormal. Benzene, a chemical of substantial environmental importance, and certain drugs have been associated with the production of certain types of leukemias in humans (Chapter 6).

Aggregation of platelets is activated by complex biochemical processes that come into play following injury to the vascular wall; the aggregated platelets stem bleeding. But *thrombosis* – the production of intravascular clots – can be life-threatening. If the clot forms on a plaque in a coronary artery, myocardial infarction (MI) is probably imminent. Aspirin and other drugs are effective as inhibitors of platelet activation, and so reduce the risk of MI and clot-induced (ischemic) strokes. Of course, in the presence of these drugs platelets cannot aggregate when they need to, so all can induce excessive bleeding as side effects.

The tiny blood vessels called capillaries run close to the individual cells that comprise the various organs of the body, but do not touch them. Instead the nutrients, oxygen, and foreign chemicals that they carry migrate into certain tissue fluids (called intercellular, or interstitial, fluid) which surround and bathe cells. This fluid provides the contact between the circulatory system and the body's drainage system, called the *lymphatics*. Intercellular fluid carries nutrients, oxygen, and chemicals to cells, and carbon dioxide, organic waste (including the chemicals or their metabolites) away from them. Some of these wastes enter the capillaries that combine to re-enter the veins, and the rest (particularly waste molecules too large to enter venous capillaries) pass into lymph.

The lymphatic system consists of vessels and various small organs. The lymph nodes (or glands) are found at several locations along the system of lymph vessels, often bundled into groups (as in the armpits). The glands produce one class of white blood cells called *lymphocytes*, cells that produce the body's "defense proteins," called antibodies. Other lymph glands include the spleen, the tonsils, and also the thymus. The last mentioned is located in the upper region of the chest and wastes away – atrophies – after puberty. Lymph glands are all capable of trapping foreign bodies such as proteins and bacteria; they are main lines of defense against infections ("swollen glands," which can be detected by feeling areas of the body where the nodes group together, are indications that the body has been invaded by infectious agents).

The liquid carried in lymph vessels is called lymph; it contains lymphocytes and substances acquired from intercellular fluids. Lymph is a key disposal system for the body's waste, including the debris from infectious agents and foreign proteins trapped within.

The lymphatics are intimately involved in the several complex processes whereby the body protects itself from foreign agents – the immune system. A rapidly evolving discipline is that called *immunotoxicology*, the study of the adverse effects of chemicals on the components and operations of the immune system. The consequences of exposure to immunotoxic agents range from suppression of immunity, which can lead to reduced resistance to infection and certain diseases including cancer, to mild allergic responses. Some important immunosuppressive agents include benzene, PCBs, and a variety of therapeutic drugs. Chemicals capable of producing allergic responses include toluene diisocyanate (TDI), a very important commercial product used to make certain plastics and resins that we mentioned above in connection with allergic responses in the respiratory system, and certain metals such as nickel. Animal models to study immunotoxicity are still under development and are not routinely used for testing new chemicals; some aspects of immune system toxicity can be detected using conventional test designs, but specialized tests are needed for a thorough evaluation. Although some forms of alteration of the immune system are clearly detrimental to health, there is still debate about the relevance to health of certain immune system changes that can be induced by chemical exposure. It is not clear that all biological changes brought about by a chemical will threaten health. There are vast areas of immunotoxicity waiting to be explored, and it is certain to become an increasingly important component of the toxicological evaluation of environmental chemicals.

## Liver – hepatotoxicity

Liver damage produced by toxic chemicals is called *hepatoxicity*, and has been under study for a very long time; this large and interesting organ has perhaps been subject to more extensive examination by toxicologists than any other. Among many others the liver plays a key role in digestion, regulation of blood sugar levels, storage of vitamins and iron, and synthesis of proteins and other essential molecules. The cells of the liver, called *hepatocytes*, are marvelously intricate and efficient chemical factories, and they contribute greatly to normal metabolism and to the metabolism of foreign chemicals.

This organ may appropriately be called a gland because it secretes a complex substance called *bile*. The term "gland" is applied to any organ that secretes substances, typically but not only hormones; another gland having a key role in digestion is the pancreas. Certain *endocrine* glands secrete hormones directly into the blood; among these are the *pituitary*, at the base of the brain, the *thyroid*, in the upper chest, the two *adrenals*, one just above each kidney, and the male and female *gonads*, known respectively as the *testes* and the *ovaries*. Some aspects of the endocrine system will be discussed below, in the section on reproductive toxicity.

Excreted bile is carried by a small duct to the uppermost part of the small intestine, called the duodenum, although some is stored in a sac just under the liver called the gall bladder. Bile is joined by secretions from the pancreas, and they combine to enhance digestion; the bile has a detergent-like action on fats, breaking them up into small droplets so they can be readily attacked by digestive enzymes. Bile is heavily pigmented because it contains waste chemicals generated by the liver, including breakdown products of hemoglobin, which is collected in the liver as red blood cells age and collapse. It is ultimately eliminated in the feces, to which it imparts its characteristic colors.

The liver is a prime target for toxicity because all chemicals received orally are carried directly to the liver by the hepatic portal vein, immediately after absorption. As mentioned, liver cells have an astounding capacity to metabolize these foreign compounds, in most instances turning them into water-soluble forms that can be readily excreted from the body (through the kidney). But this detoxification capacity of the liver can sometimes be overwhelmed. Moreover, some forms of metabolic change, as illustrated in the case of bromobenzene in Chapter 2, create metabolites having toxic properties more threatening than those of the original chemical. In either of the last two conditions liver damage can occur, as can damage at other sites of the body when toxic molecules escape the liver.

Toxicologists classify hepatic toxicants according to the type of injuries they produce. Some cause accumulation of excessive and potentially dangerous amounts of *lipids* (fats). Others can kill liver cells; they cause cell *necrosis*. *Cholestasis*, which is decreased secretion of bile leading to jaundice (accumulation of gruesome looking pigments that impart a yellowish color to the skin and eyes) can be produced as side effects of several therapeutic agents. *Cirrhosis*, a chronic change characterized by the deposition of connective tissue fibers, can be brought about after chronic response to several substances. And, as will be reviewed in the next chapter, liver cells can be altered by chemicals and develop into tumors, of both benign and malignant nature. Experimentalists who study the liver's many and varied responses to chemicals will caution the reader that "hepatotoxicity" is not a very helpful term, because it fails to convey the fact that several quite distinct types of hepatic injury can be induced by chemical exposures and that, for each, different underlying mechanisms are at work. In fact, this situation exists for all targets, not only the liver. Lipid accumulation – fatty livers – can result, for example, if a hepatotoxic chemical somehow alters biochemical pathways to produce an oversupply of the chemicals out of which fats (lipids) are synthesized. Another chemical can interfere with the process that normally breaks down liver fats, with the same result – lipid accumulation. That chemicals can cause fatty liver was first discovered more than a century ago, when worker exposure to yellow phosphorus, which was used to manufacture match heads, was found to be associated with this condition.

Carbon tetrachloride ($CCl_4$), once a very widely used solvent, has perhaps been the subject of more experimental study than any other organic chemical. Since the early 1920s experimentalists have been investigating its various effects on the liver and have come to understand in great detail how this molecule performs its deeds.

The carbon tetrachloride molecule has the simple chemical structure shown on the left; four atoms of chlorine are chemically bonded to one carbon atom.

Carbon tetrachloride       Trichloromethyl radical

When molecules reach the liver some are acted upon by components of cells, which manage to break apart one of the carbon–chlorine bonds. The chemical bond consists of a pair of interacting electrons (the little dots), one contributed by chlorine, the other by carbon. The bond-breaking results in the release of an atom of chlorine and a very reactive chemical group called the trichloromethyl ($CCl_3$) radical, as shown in the chemical equation. It appears that damage to the liver cells is initiated by the trichloromethyl radical, which has the capacity to interact chemically with some of the normal protein and fat molecules of the cell. These disturbances result, among other things, in the abnormal oxidation of the fats of the liver cell walls; these molecular events can result in a sequence of additional damaging events, the nature and extent of which depend upon the concentration of $CCl_3$ radicals generated and their lasting power within cells. Cell death – necrosis – can result if damage to cell walls is extensive. Carbon tetrachloride can act through other mechanisms and cause other types of liver cell damage. Some other chemicals, the closely related solvent chloroform ($CHCl_3$) for example, cause liver cell necrosis in ways similar to carbon tetrachloride, while others act in quite different ways. So we may list many chemicals capable of causing specific types of liver injury, but such listing may obscure the fact that many differing underlying molecular and cellular events may be at work.

Liver cirrhosis can result from chronic exposure to several chemicals, including carbon tetrachloride, alcohol, and aflatoxin (the mold product described in the Prologue). Over time continuous liver injury leads to cirrhosis, the accumulation of abnormal fibers made of collagen, the normal protein component of bones. Liver cirrhosis following long-term ingestion of excessive amounts of alcohol presents an interesting toxicological problem. Some scientists believe cirrhosis is a direct consequence of alcohol-induced toxicity. But others believe the evidence points to a kind of indirect effect – specifically, that the cirrhosis actually results from chronic nutritional deficiency frequently associated with alcoholism.

Several clinical chemical tests are available to detect the presence of liver injury. Certain normal liver enzymes can be released to the blood following injury to the cells containing them, and a search for their presence is a routine component of chemical testing and of monitoring of animals during toxicity testing. Other tests, routinely performed during animal testing and on human beings subjected to medical examination, provide information about the nature and extent of hepatic disease.

### Kidney – nephrotoxicity

The kidneys – a pair of organs in the lower back region, just below the ribs – are part of the *urinary system*, the main function of which is to rid the body of waste substances, including those resulting from normal biochemical processes and those resulting from absorption of other, non-essential chemicals. The kidneys are essentially filters, removing wastes from blood carried through them. They also play a critical role in regulating the contents of body fluids. Urine is formed in the kidneys and is carried to the *bladder* by two tubes called ureters. The urinary bladder stores urine until the volume reaches a certain level, whereupon it is released through a tube called the *urethra* to be excreted from the body.

The main filtering units of the kidneys are called *nephrons*; about one million nephrons are present in each kidney. Each nephron consists of a *renal corpuscle* and a unit called a *tubule*. Blood carrying normal metabolic wastes such as urea and creatine moves through a portion of the corpuscle called the *glomerulus*, where a filtrate forms that contains water, normal metabolic products, and also waste products; the filtrate collects in another unit called Bowman's capsule. Glomerular filtrate then moves into a highly convoluted and multifaceted set of tubes – the tubule – where most useful products (water, vitamins, some minerals, glucose, amino acids) are taken back into the blood, and from which waste products are collected as urine. The relative amounts of water and minerals secreted or returned to the blood are under hormonal control.

Chemicals toxic to the kidneys can injure different components of them and thereby undermine their function in several ways. Several so-called "heavy" metals, notably mercury, cadmium, chromium, and lead, are particularly ruinous to the tubules. Certain concentrations of these metals present in glomerular filtrate can seriously impair the functions of the tubules, and this can lead to loss from the body of excessive amounts of essential molecules such as sugar (glucose) and amino acids. If concentrations of these metals reach high enough levels cell death can follow, which, if extensive enough, can close down kidney function altogether. The two hepatoxicants mentioned earlier, carbon tetrachloride and chloroform, also cause nephrotoxicity.

Certain antibiotics such as the tetracyclines, streptomycin, neomycin and kanamycin can cripple the tubules if taken in excessive amounts. Toxic damage to the kidneys can affect not only their filtration functions, but can alter the organs' control over blood levels of certain critical molecules. A complex biochemical–hormonal system controlling blood pressure and volume, for example, is regulated by the kidneys, so that chronic kidney damage can inflict damage on the circulatory system, including the heart itself. The kidneys also exert control over molecules critical to hemoglobin synthesis.

A number of clinical tests are available to detect kidney damage. The clinician examining a patient or the toxicologist monitoring an animal toxicity study collects urine and blood samples. Indications of kidney damage (which, of course, for the human patient could be related to many factors other then chemical toxicity) include urinary excretion of excessive amounts of proteins and glucose, and excessive levels in the blood of unexcreted waste products such as urea and creatine. A number of additional kidney function tests are available to help pin down the location of kidney dysfunction.

### Nervous system – neurotoxicity

The nervous system consists of two main units: the *central nervous system* (CNS), which includes the *brain* and the *spinal cord*; and the *peripheral nervous system* (PNS), which includes the body's system of nerves that control the muscles (motor function), the senses (the sensory nerves), and which are involved in other critical control functions. The individual units of the nervous system are the nerve cells, called *neurons*. Neurons are a unique type of cell because they have the capacity to transmit electrical messages around the body. Messages pass from one neuron to the next in a structure called a *synapse*. Electric impulses moving along a branch of the neuron called the *axon* reach the synapse (a space between neurons) and cause the release of certain chemicals called *neurotransmitters*, one of which, acetylcholine, we described earlier in the chapter. These chemicals migrate to a unit of the next neuron called the dendrites, where their presence causes the build-up of an electrical impulse in the second neuron.

There are three types of neuron. *Sensory* (or afferent) neurons, as their name implies, carry "sense" information about the body; *motor* (or efferent) nerve cells carry "instructions" from the CNS, in the form of nervous impulses, to the muscles, organs, and glands of the body. The *associated neurons*, which have several other names, are involved in detecting impulses from sensory neurons and passing them to motor neurons.

Nerves are bundles of nerve fibers, connective tissue, and blood vessels. Each fiber is part of a neuron. Afferent nerves are entirely comprised of sensory neurons and efferent (motor) nerves are comprised of motor neurons.

The nervous system has many complex components but we shall not go more deeply into its structure and functioning here. Obviously, toxic damage to certain of its components can be unfavorable for the whole organism, because the nervous system is intimately involved in control of virtually all of the body's mental and physical functions.

Two additional features of the nervous system are particularly important to a discussion of neurotoxicity. The *blood–brain barrier* and the *blood–nerve barrier* act as protective devices for the nervous system, and are effective at preventing movement of certain chemicals from the blood to the brain and nerves. Unfortunately neither barrier is effective against all types of molecule, and there are plenty of examples of brain and nervous system toxicants that can penetrate the "barriers."

Neurotoxicologists generally categorize effects according to the chemical's primary site of action.

Some substances, such as carbon monoxide and barbiturates, can deprive brain cells of oxygen or glucose – they produce anoxia – with potentially serious consequences for gray matter. Other substances, such as lead, hexachlorophene, and the antitubercular drug isoniazid, are capable of causing loss of myelin, a coating or sheath for the axon and dendrites that extend from the central unit (cell body) of neurons. Demyelination can occur in either the CNS or PNS.

So-called *peripheral neuropathies* can result from excessive exposure to certain industrial solvents such as carbon disulfide ($CS_2$, used in the rubber and rayon industries) and hexane ($C_6H_{14}$, once used in certain glues and cleaning fluids). Over-exposure to acrylamide, an important industrial chemical, and chronic alcohol abuse can also induce this effect. As the name implies, it involves attack of the chemical on and damage to axonal portions of neurons. Typical symptoms of peripheral neuropathies include weakness or numbness in the limbs, which are more or less reversible depending upon the specific agent and the intensity of exposure.

A particularly interesting example of the role of metabolism in toxicity was brought to light during the 1970s by Herbert Schaumburg and Peter Spencer. We have mentioned that the solvent hexane (structure I) is capable of causing delayed peripheral neuropathy in the form of axonal degeneration. Excessive occupational exposure to hexane leads to loss of sensation in the limbs, particularly in the hands and feet. Muscle weakness in the extremities is common. The effects are usually delayed, and occur several months to a year following initial exposure.

The active toxicant is not hexane, but the hexane metabolite called 2,5-hexanedione (II):

Hexane (I)

2,5-Hexanedione (II)

Methyl-*n*-butyl ketone (III)

The latter compound results from loss of two hydrogen atoms by metabolic oxidation at each of two of the carbon atoms of hexane, and their replacement by oxygen atoms. Schaumburg and Spencer not only demonstrated this, but also showed that the identical neurotoxic events result from direct exposure to compound II and also another common solvent called methyl-*n*-butyl ketone (III). Chemical III is, like hexane, readily metabolized to the active toxicant, molecule II. Because both I and III yield the same metabolite (II), and because this metabolite is the source of toxicity, then exposure to both of these chemicals produces the identical type of neuropathy.

Certain organic forms of mercury can elicit specific damage in the main cell body of peripheral neurons. Similar responses are associated with certain natural products called vincristine and vinblastine, both of which have been used as antileukemic medicines. The deadly botulinum toxins, mentioned earlier in this chapter, block transmission of nerve impulses at the synapses of motor neurons. This blockage results in muscular paralysis which, if sufficiently severe, can lead to death, usually because respiration is impaired. The once widely used pesticide, DDT, is an organic chemical that also acts on the nervous system at this site, although it can also mount an attack on areas of the CNS.

The CNS is a target for many neurotoxicants, including mercury, certain forms of gold, manganese, and even the food ingredient monosodium glutamate (MSG), although very large doses of the latter are required. Depending upon their location and severity, effects on the CNS can have profound consequences for sensory and motor functions. Chronic manganese intoxication, for example, produces signs and symptoms that in advanced stages mimic the condition known as Parkinsonism. The most common early indications of chronic poisoning with this metal include incoordination, difficulty in keeping a steady gait, impaired speech, and severe tiredness. Tremors, weakness in the limbs leading to an inability to stand upright, and even emotional disturbances appear at more advanced stages. Severe tremors during resting is a particularly telling characteristic of manganese poisoning. Removal of the worker from manganese environments and perhaps treatment with agents that assist removal of manganese from the body are necessary to alleviate this severe condition. There is considerable ongoing study and debate regarding the minimum exposure level needed to produce manganism.

Excessive exposure to inorganic mercury, particularly in its elemental form, creates a psychological condition called erethism. Victims suffer from excessive timidity and self-consciousness, inability to concentrate, loss of memory, and other psychological changes. From at least the seventeenth and well into the nineteenth century, mercury was used to cure felt, and workers exposed during that process could acquire erethism. Lewis Carroll's character the Mad Hatter was no doubt based on the fact that hatters exposed to mercury could in fact go mad. The phrase "mad as a hatter" was in common use at the time *Alice's Adventures in Wonderland* was written.

A rather insidious form of mercury neurotoxicity can occur when the element is bound chemically to certain groupings of organic molecules – so called alkylmercury compounds. Such forms of mercury, unlike the inorganic forms, can cross the blood–brain barrier (which is fairly effective at excluding inorganic mercury) and damage brain cells in serious ways. The most devastating manifestation of this form of neurotoxicity developed at Minamata, Japan, during the 1950s and 1960s, and perhaps continued to the mid 1970s. Discharge of elemental mercury from a chemical plant into Minamata Bay created high levels of the element in the water. When in this environment, the mercury can be converted to the organic form, called methylmercury, by microorganisms present in sediments. The organic mercury compound found its way into fish and shellfish, and thence to humans, mostly local fishermen and their families. The disease was first identified as organic mercury poisoning in 1963. The patients exhibited visual and hearing impairments, and mental disturbances characterized by periods of agitation alternating with periods of stupor. Similar outbreaks of organic mercury poisoning occurred in Iraq in 1956, 1960, and 1971; in these cases the source was grain that had been treated with an organic mercury compound to prevent mold growth. The grain was supposed to be used only for seed, but was instead taken for grinding to flour. The 1971 outbreak affected more than 6500 people, of whom more than 400 died.

The particular form of mercury that accumulates in fish or shell-fish, methylmercury, has been the subject of extensive investigations in recent years, and results from these studies tell us much about the potential some chemicals have for interfering with the highly sensitive processes that are at work to build the nervous system during the developmental period of life. We shall come back to this subject later, when the subject of developmental toxicity is covered.

Perhaps the most important of the neurotoxic agents found in the environment is lead (chemical symbol, Pb, from the Latin word for lead, plumbum).

People have been extracting lead from ores for about 4000 years. Its uses have been many and have changed over the centuries, sometimes because of concerns about the metal's danger to health. Lead has probably been the subject of more toxicological investigations than any other substance. An unusually high proportion of these investigations involve observations in humans. There are two reasons for this: (1) human exposure to this metal has been and still is very widespread; and (2) a means has been available for some time to obtain a direct measure of that exposure. The latter is the concentration of lead in human blood, commonly expressed as micrograms Pb per deciliter (100 ml) of blood ($\mu$g/dl). It is relatively easy to obtain this value by taking a sample of blood and subjecting it to chemical analysis.

Young children are especially susceptible to the effects of environmental lead, first because their bodies accumulate lead more readily than do those of adults and, second, because they appear to be more vulnerable to certain of the biological effects of lead. In 1988 the US Public Health Service estimated that, in the United States alone, 12 million children were exposed to leaded paint, 5.6 million to leaded gasoline, 5.9–11 million to dusts and soils containing excessive lead, 10.4 million to lead in water (in part because of lead in pipe solders) and 1.0 million to lead in food. The Public Health Service also estimated at that time that nearly 2.4 million children in the United States exhibited blood lead values greater than 15 μg/dl, and that nearly 200 000 children exhibited levels greater than that at which some form of medical intervention was appropriate.

Substantial progress in reducing lead body burdens has been made in the past three decades (e.g., by programs to eliminate leaded gasoline and to prevent children's exposure to leaded paints present in many older homes), and the most recent nationwide surveys have revealed that average blood levels in children are now near 2 μg/dl. There are still subpopulations showing excessive exposure, probably related to paint found in older homes. In some areas of the world, particularly where leaded gasoline is still used, the problem remains serious. The problem is complicated by the fact that, as we have worked to reduce lead exposures, health scientists have uncovered new concerns.

Lead has multiple toxic effects, on elements of the blood, where it interferes with heme synthesis, and on the kidney. But of greatest concern are the effects of lead on the central nervous system. Encephalopathy – brain disease – and peripheral neuropathy occur when blood levels reach the range of 70–100 μg/dl, a level which is nowadays rare in children, at least in the developed world. Lead encephalopathy manifests itself as stupor, coma, and convulsions, and is generally accompanied by severe cerebral edema and other types of brain cell damage. Until the 1980s a blood lead level of 25 μg/dl or less was thought to be without much medical significance, and was sufficiently low to prevent encephalopathy, peripheral neuropathy and even other, less serious manifestations of neurologic injury. But beginning in 1979, when Herbert L. Needleman and his associates at the University of Pittsburgh School of Medicine reported so-called "subclinical effects" at blood lead levels below 25 μg/dl, and continuing over the following 25 years as Needleman, and other investigators as well, uncovered such effects at lower and lower blood lead levels, public health authorities came to discredit the 25 μg/dl limit.

What are these "subclinical effects?" Very simply, they are effects that occur at blood lead levels below those that produce clinically measurable effects – they occur in the absence of any sign of overt lead poisoning. These effects can be detected only by studying various forms of behavior, such as degree of hyperactivity and classroom attention span, and performance on various tests of intelligence and mental development. Deficits in neurobehavioral development, as measured by two widely used tests – the Bayley and McCarthy Scales – have been reported in children exposed prenatally (via maternal blood) to blood levels in the range of 10–15 μg/dl. Small IQ deficits have also been reported at even lower blood lead levels, and some experts suggest the threshold for these effects, if there is one, has not been identified.

Needleman and others who are studying this problem observe small shifts in, among other things, the distribution of IQ in populations of children exhibiting blood lead levels in a range below that known to produce clinical effects. Are such shifts of medical or social importance? For the individual child a small downward shift in IQ, for example, may not be of significance, especially if it does not persist beyond early childhood. But for society as a whole such a shift may be of substantial significance if millions of children are included, and the effect persists into later life. This is a considerable dilemma that public health officials have not yet fully faced, although it is clear the trend in public health and regulatory policy is to move toward the goal of reducing blood lead levels. At present most authorities define lead exposure as excessive when blood levels exceed 10 μg/dl; some experts are now calling for stricter standards.

Signs of neurotoxicity can be detected in several ways. Assessment of mental state and sensory function can be performed in humans; in fact it is difficult or impossible to make such assessments in experimental animals (you don't get much of a response from a rat when you ask if he's often feeling tired, has recurring headaches, or is having difficulty seeing). There are objective tests of sensory function, particularly as it involves the optic and auditory nerves; responses to light and sound stimuli can be readily measured. Motor examination includes inspection of muscles for weakness and other signs of dysfunction, such as tremors. Specific types of motor disorders give an indication of the part of the nervous system that has been affected.

Reflex action can be objectively examined. Measurement of the so-called electrical conduction velocity of motor nerves gives an indication of whether peripheral nerves have been damaged. Electromyography, or examination of the electrical activity of the muscle, and electroencephalography (EEG) can also be used to detect the presence of neurologic abnormalities. And, at least in animal tests, detailed pathological examination of neural tissue can be performed on animals that are killed.

A particularly rapidly growing area of the science is *behavioral toxicology*. The nervous system serves a vast integrative function for the body, and various forms of behavior can be altered by response to substances that affect the nervous system in some way. We have already seen some examples of behavioral changes associated with exposure to lead, mercury, and manganese. The incorporation of behavioral measures into animal tests is becoming increasingly common.

### Reproductive and developmental toxicity

The reproductive systems of both males and females can be harmed by particular chemicals. In males certain chemicals cause the testes to atrophy and reduce or eliminate their capacity to produce sperm. Particularly striking in this regard is a now banned but once widely used pesticide called DBCP, residues of which persist in ground water supplies in a few regions of the country. Its pronounced impact on spermatogenesis is readily detectable in experimental animals and, unfortunately, has also been observed in some men once occupationally exposed to large amounts. The heavy metal cadmium is another substance effective at reducing sperm production.

Impairment of the reproductive process can occur in several other ways, generally by the induction of abnormal physiological and biochemical changes that reduce fertility, prevent the full maturation of the fetus, or prevent successful birth (parturition). Reproductive success, or lack thereof, is measured in animal experiments in which mating is allowed to occur between chemically treated females and untreated males, or vice versa, or between males and females that are both exposed to the chemical. Various indices of reproductive performance are then measured. The fertility index, for example, is the percentage of matings that result in pregnancy. The gestation index is the percentage of pregnancies resulting in the birth of live litters. The percentage of offspring that survive four days or longer is called the viability index. These and several other indices are the measures by which treated and control animals are compared.

An area of emerging importance is *developmental toxicity*. Here we are concerned about the effect of the chemical on the developing embryo and fetus (exposures received in utero), and on the further development of the infant and child subsequent to birth, at which time chemical exposure may cease, or may continue to weaning because the chemical received by the mother is transferred to her milk. Or it may continue throughout life because there are sources of the chemical in addition to that supplied by the mother because of her exposure.

A particularly important type of developmental toxicity is called *teratogenicity*. After fertilization the ovum – a single cell – begins to proliferate, making more of its own kind by a series of divisions. In humans, at about the ninth day the remarkable process of cell differentiation begins; the specific types of cells (e.g., neurons, liver cells, etc.) that make up the body begin to form and to migrate to their appropriate positions. This *embryonic* period lasts until the fourteenth week of gestation in humans. This is the period of *organogenesis*, which means exactly what it appears to mean – the various organs of the body are generated, although at different times and rates. Following the embryonic is the *fetal* period, during which the organism grows and bodily functions mature. The fetal stage ends at birth, although obviously development continues to take place following this event. Although the total duration of embryonic and fetal development varies among species, the sequence and relative timing of the critical events is about the same in all mammals.

Thalidomide, the chemical structure of which is shown, was introduced in 1956 by a German pharmaceutical firm for use as a sedative. The drug was widely used, though not in the United States, at oral doses of 50–200 mg/day to reduce nausea and vomiting during pregnancy, and it was quite effective at doing so. Few side effects were experienced by women taking the drug.

An unexpected increase in the incidence of certain rare and devastating birth defects – absence of limbs and reduced limb length – was reported beginning in 1960, first in West Germany and then in other areas of the world. Thalidomide was identified as the cause through the work of W. G. McBride in Australia and W. Lenz in West Germany.

Thalidomide

The drug was taken off the market in 1961 and the increased incidence of these rare birth defects disappeared at the predicted time thereafter. Very few cases were reported in the United States, where full-scale entry into the marketplace was delayed because of some careful work by Dr. Frances Kelsey of the FDA. Perhaps 8000–10 000 cases were reported worldwide. The offspring of women taking the drug during the sixth and seventh weeks of pregnancy were found to be at highest risk of developing birth defects. This is the period of organogenesis in which the skeleton is under most rapid development.

Thalidomide is a *teratogen*, in humans and in experimental animals, although the latter were found to be susceptible to these effects only after an extensive set of investigations was undertaken following the tragic human findings. "Teras" is the Greek word for "monsters." That some humans are born with structural abnormalities of the body – defects of the palate, of the skeletal system, of the heart, the eye, and so on – has been known since antiquity, and the appearance of human "monsters" of one sort or another in legend, myth, and in actuality is a prominent feature of human history. Clearly teratogens and other factors that cause birth defects have been around for a very long time and predate the chemical revolution by at least several millennia.

The more common medical term for birth defects is "congenital abnormalities," meaning those that are inherited. The earliest medical hypotheses suggested a genetic cause for these abnormalities, not an environmental one. This is probably still true for many such abnormalities, but in the 1930s it was learned that terata could be induced in animal species by manipulating their environment – in this case by withholding adequate amounts of riboflavin or vitamin A from pregnant females. Since that time several environmental agents have been found be capable of causing "congenital" abnormalities in humans, and these and a great many more have been found to be animal teratogens. It is also noteworthy that livestock and other domesticated animals as well as wild animals are all susceptible to environmental teratogens, and some of the most potent teratogens occur as natural components of certain range weeds that are sometimes consumed by cattle or sheep.

Although there are some chemicals that can cause frank birth defects, other effects on the developing organisms resulting from in utero exposure and from exposure in the *neonatal* period, are also of interest. Thus, the broader term *developmental toxicity* has come into use, and refers to any adverse effect on the developing organism, from reduced birth weight, to impaired neurological development, to frank birth defects or death.

Experimental studies reveal quite conclusively that the timing of exposure to a developmental toxicant – the time during gestation the dose is received – is as important as the size of the dose. Because the rate and timing of cell differentiation and organogenesis vary among organs, they will be differently affected depending upon the period of dosing. In the rat, for example, dosing early in the period of organogenesis tends to cause effects on the eye and brain much more frequently than defects of the skeleton, or the urinary system, or the palate, while the opposite is true for dosing during the mid-point of that period. A few days can make a large difference in the severity produced by, and target for, a developmental toxicant.

Typical animal tests for developmental toxicity involve rats and rabbits, and sometimes other species. The agent to be tested is administered to groups of pregnant animals at several dose levels during the entire period of organogenesis (days 6–15 of gestation in the rat). Fetuses are usually removed from the mother one day prior to expected delivery, and various examinations are made to determine the fate and health status of each fetus and to identify physical abnormalities. Because dosing occurs throughout organogenesis these tests can capture a broad range of possible developmental effects.

A particularly important aspect of the developmental toxicity study concerns the effect of the chemical on the treated mother. If a dose that is in some way toxic to the mother is used in these tests – and the highest dose is often toxic in this way – the test results may not be telling with respect to the offspring. Thus, if the offspring from mothers that experience a toxic response are also affected, it will not be clear whether the chemical is itself developmentally toxic, or whether the health of the fetus was injured only because of its mother's unhealthy condition. Generally, toxicologists do not consider an agent a developmental toxicant if its effects on offspring appear only at the "maternally toxic dose," and not at lower doses.

Toxicologists nowadays take a broad view of developmental toxicity; they consider not only *structural* but also *functional* abnormalities to qualify as adverse, as long as they were produced as a result of exposures incurred in utero. Thus, for example, the developmental effects of chronic alcohol abuse by pregnant women, known as *fetal alcohol syndrome* (FAS), are characterized not only by the presence of certain craniofacial abnormalities, but also by a variety of disabilities such as shortened attention span, speech disorders, and restlessness. Although fully expressed physical deformities included in FAS are associated with heavy drinking, debate continues on the level of alcohol consumption, if any, that is without these more subtle effects on behavior.

Public health officials strongly discourage the consumption of any amount of alcoholic beverages by pregnant women.

Another feature of developmental toxicity is raised by the experience we have had with the drug DES – diethylstilbestrol. This is an easily synthesized chemical having potent *estrogenic* properties – its chemical characteristics are such that it can fulfill many of the biological functions of the normal female sex hormones, or estrogens. From the mid 1940s until 1970, the drug was widely used in pregnant women who had abnormally low levels of estrogen production, to prevent threatened miscarriage. In the late 1960s, Arthur Herbst of the Massachusetts General Hospital reported an unusual occurrence of a cancerous condition called clear-cell adenocarcinoma of the vagina. What made the occurrence unusual was the fact that the seven cases Herbst saw in the 1966–1969 period were in women between the ages of 15 and 22. This particular cancer was extremely rare in this age group.

A series of investigations initiated during this period led to the conclusion that DES use by the mothers of these young women was the causative factor. Diethylstilbestrol is a developmental toxicant, because its effect was the result of exposures received in utero, but that effect – a rare form of vaginal cancer – was not fully expressed for about two decades following the exposure (the age at which the incidence of the disease peaked was later shown to be 19). The DES produced not only cancer but abnormalities of the sexual organs in both male and female offspring, most of which were not expressed until the children passed the age of puberty.

Diethylstilbestrol is the leading member of a class of toxicants that is now under intense study. The term *endocrine-disruptors* is often used to describe the class. It is meant to describe any compound that causes adverse effects by somehow interfering with the functioning of any element of the endocrine system. Some compounds may do this because their structures mimic those of the natural hormones (DES has broad structural resemblance to the sex hormone estradiol), and thereby "fool" the cells where these hormones normally act. The reproductive system operates, of course, under a complicated set of endocrine controls, so it is a leading focus for the study of endocrine disruptors. The thyroid gland excretes hormones that have multiple and critical roles in regulating the body's metabolic functioning. The controversial environmental contaminant perchlorate can, at relatively high doses, alter thyroid function in adverse ways. More discussion of the endocrine disruption issue is reserved for Chapter 9.

Methylmercury, which we referred to in the neurotoxicity section, occurs in fish and shellfish found in both the ocean and fresh water systems. The mercury that is the source of methylmercury arises from power plant emissions and industrial processes. Some even comes from dental amalgam waste and from natural sources in the ocean bed. It is a developmental toxicant, causing severe defects (recall the Minimata Bay story), but also causing behavioral and learning impairments in the offspring of women exposed during pregnancy.

Several recent epidemiological studies have involved examination of populations that consume unusually high levels of fish. One of these, conducted in the islands of the Seychelles, has not so far revealed behavioral and learning impairments in children whose mothers exhibited mercury levels (measured in hair) higher than those typically seen in the United States and European countries. But another study, conducted in the Faroe Islands, turned up evidence of cognitive and behavioral impairments in children. Scientists have struggled to understand why two well-done studies have produced such different outcomes, and some possible reasons have been suggested. The EPA and public health officials have acted on the basis of the Faroe data, out of both caution and also because they seem to be supported by other, more limited data, and by experimental studies. The debate is not so much about whether methylmercury is a developmental toxicant, but rather over the dose required.

This particular issue of developmental toxicity is complicated in a most interesting way. Fish are a critical source of certain fatty acids that are important contributors to normal neurological development. So, should we ask women who are pregnant to reduce their intake of a food that is important to successful neurological development, so as to reduce their intake of a chemical that may adversely affect those same developmental processes? This is a risk management problem with no easy solution. The FDA and EPA have jointly issued guidelines on fish consumption during pregnancy, and they seem to reflect this concern for adequate nutrition during this critical period (more on this in Chapter 11).

It should not be assumed that all birth defects and functional abnormalities in children are caused by drugs or environmental chemicals. It is clear that environmental factors such as extreme heat or cold, certain forms of radiation, infections (particularly German measles and syphilis), dietary deficiencies, and genetic disorders in the parents can all put the developing fetus at risk.

Because our intention in this book is to emphasize principles and not toxic agents, we shall here complete the survey of toxic agents and their targets. Although many of the examples presented are environmental chemicals of considerable importance (e.g., CO, Pb, Hg), the examples were chosen primarily to illustrate certain principles of toxicity: the importance of dose and exposure duration, of the chemical form of the toxic agent, of exposure route, and especially in the case of developmental toxicants, of the timing of exposure. They also illustrate the role of ADME in the production of toxicity and the fact that toxicity targets are many and varied; in fact, we have seen that chemicals exert their effects in many different ways on the same target – not all substances listed as liver toxicants act in the same way. Our survey is, however, not complete and by no means reflects all that toxicologists know.

Before we plunge into the world of carcinogens, we should note that all of the toxic phenomena we have described exhibit dose–response relationships and that LOAELs and NOAELs can be identified for all. As we shall see in later chapters these quantitative features of toxic phenomena are at center stage when we begin to examine risk to exposed populations.

# 5

# Carcinogens

People all over the world are exposed to cancer-causing chemicals present in air, water, food, consumer products, and even in soils and dusts. In their places of work some people come into contact with additional cancer-causing agents, generally at higher exposure levels than those experienced by the general population. Some people deliberately expose themselves, and incidentally expose others, to the large number of known and suspected carcinogens present in tobacco smoke. People are also exposed to various physical agents – ultraviolet radiation from the sun and sunlamps and other forms of natural and artificially produced radiation – that increase cancer risks. We are all being assaulted by chemical and physical carcinogens. Add to this the substantial viral and genetic contributions. No wonder the chances of developing some form of cancer over our lifetime is about one in three (for women) and one in two (for men).

But we are moving too quickly. Before we can begin to contemplate the contribution of all these environmental carcinogens to the total cancer problem we need to acquire a better understanding of what is meant by the terms "carcinogen" or "cancer-causing chemical" and of how certain substances get to carry these labels.

We shall begin with a little history, and then move to a discussion of cancer statistics and the causes of cancer, and then provide some background on cancer biology and the mechanisms of tumor development. Some of the general characteristics of chemical carcinogens will also be covered. The methods for identifying chemical carcinogens are the subject of Chapter 6. How their risks are estimated is left to later chapters.

# Cancer and chemical carcinogens – historical perspective

The large group of diseases we refer to as cancers[1] have in common cells that have lost the capacity to control their own growth. This disease can arise in any organ or tissue of the body. The unregulated growth of body cells results in the production of masses that compress, invade, and destroy contiguous normal tissues. Cancer cells then break off or leave the original mass and are carried by the blood or lymph to distant sites of the body. There they set up secondary colonies, or metastases, further invading and destroying other organs.

It is certain that these diseases are not totally a product of the industrial age or the era of modern chemical technology. Lack of solid statistical information forces us to avoid the question of how much human cancer there would now be if the industrial revolution and its chemical and physical products had never appeared on earth, but if the average age at death had nevertheless increased exactly as it has over the past two centuries. This is, of course, an unlikely historical scenario, in that products of the industrial revolution have contributed substantially to the fact that more people are living to old age. Particularly important have been medicines that prevent death in early childhood, antibiotics that cure infections, agricultural technology, and many forms of medical technology and public sanitation. We also bring up the issue of average age at death because most cancers are diseases of old age. If many people die early from other diseases, then the numbers of people alive to contract cancer are fewer. To be meaningful, statistics on cancer rates must contain an adjustment for differences in the distribution of ages in the populations under investigation, whether comparisons are being made for the same population at different points in time, or for different populations at the same point in time.

Human cancers were much discussed by Galen and most medical commentators ever since, and dozens of hypotheses regarding the origins (*etiologies*) of these diseases are recorded in the medical literature. A seminal event relevant to our present concerns about the environment occurred in 1775. A British surgeon, Percival Pott, published his observations on high rates of cancer of the scrotum among London chimney sweeps. Pott attributed the cancers to the soot with

---

[1] Early Greek medicine recognized cancers, and tumors were described by Galen, the Greek physician to Roman Emperors (second century AD), as "crab-like" in form. *Karinos*, the Greek word for crab, is the origin of the English word cancer. The Latin word for crab is *cancer*.

which these workers came into contact. The surgeon reached conclusions about the causal relationship between soot and scrotal cancer on several grounds, not least of which was the fact that the occurrence of these cancers could be reduced if certain hygienic practices were followed to reduce direct contact with soot. Pott's observations can be said to be the first to reliably establish a cause–effect relation between an environmental agent and cancer, and also to recognize the importance of good industrial hygiene measures to protect workers from hazardous agents.

For the 16 centuries prior to Pott's observations the medical view of cancer was the one proffered by Galen in about 200 AD. The Greek physician drew upon then-current views of the composition of the human body and postulated that cancers were caused by certain imbalances within the body. The four body "humors" (the biological counterparts of the famous basic elements of nature – Earth, Air, Fire, and Water) were blood, phlegm, yellow bile, and black bile. Diseases resulted from imbalances in these humors, and excessive amounts of black bile (*melancholia*) gave rise to cancer. That melancholy states contributed to cancer dominated medical thinking until well into the nineteenth century, and it is of course still fashionable in some circles. Pott's observations prompted other investigations into external sources of the problem, and a series of similar observations throughout the nineteenth century, involving other sources of "soots and tars," arsenicals used in medicine, and occupational cancers in the developing chemical industry, gradually undermined the Galenian theory.

Pott knew nothing about the chemical composition of soot; in fact, his paper does not say much about causal agents or the possible importance of the findings. We now know that soots are mostly composed of inorganic carbon (a biologically rather inert material[2] which is also the major ingredient of graphite and, in crystalline form, of diamond), but they also contain small amounts of many different chemicals that are grouped under the general heading of polycyclic aromatic hydrocarbons (PAHs). The PAHs occur as degradation products whenever any organic materials – fuels, foods, tobacco, for example – are burned or heated to high temperature. These chemicals are also present in unburned petroleum and products such as coal tars. Occupational skin cancers associated with materials related to soots were reported

---

[2] Carbon is chemically rather inert, but fine particles of carbon are among the categories of Particulate Matter discussed in the previous chapter.

by several investigators in England and Scotland during the last quarter of the nineteenth century. Ross and Cropper, two British scientists, proposed in 1912 that coal-tar related cancers were induced by chemicals, the same chemicals also found in soots and combustion products of various sorts.

Although they did not know it at the time, two Japanese scientists, Katsusaburo Yamagiwa and Koichi Ichikawa, provided in 1915 indirect but nevertheless significant experimental confirmation of Pott's observations on soots, when they were able to produce skin tumors on the ears of rabbits to which they had applied coal tar (not soot) for many months. The work of the Japanese investigators is also important because it represented the first laboratory production of tumors with an environmental chemical (or chemical mixture).[3] Of particular interest was their observation that the tumors (which they called "folliculoepitheliomata") appeared only after many months of continuous application of the cancer-causing agent. That studies of chronic duration are necessary to detect most carcinogens has been amply confirmed since the pioneering work of Ichikawa and Yamagiwa.

Yamagiwa was justifiably proud of his achievement and wrote a haiku, perhaps the only one "celebrating" this dreaded disease.

> *Cancer was*
> *produced!*
> *proudly I walk*
> *a few*
> *steps.*

In the 1920s the British scientist, Ernest L. Kennaway (1881–1958), suspecting that the carcinogenically active components of coal tar were to be found among the PAHs, tested one member of the class called dibenz[a,h]anthracene on the skin of shaved mice and found it to be carcinogenic. This work, reported in the *British Medical Journal* in 1930 (Kennaway and Heiger), was the first in which a single chemical compound was shown to be capable of producing tumors. A team of chemists at the London Free Cancer Hospital processed about two tons of coal tar pitch and isolated small amounts of a pair of isomeric PAHs, benzo(a)pyrene and benzo(e)pyrene. The former proved to be

---

[3] Certain animal cancers had already been shown to be produced by viruses by Ellerman and Bang (1908) and Rous (1911).

carcinogenic, the latter not. The chemical structures of these two PAHs and Kennaway and Heiger's PAH are as shown, in a shorthand form:

Dibenz[a,h]anthracene

Benzo[e]pyrene

Benzo[a]pyrene
(long form)

Benzo[a]pyrene
(abbreviated form)

These chemical structures are simply abbreviated forms of those introduced in Chapter 1. Polycyclic aromatic hydrocarbons are composed of carbon and hydrogen atoms only. The form of the abbreviation is illustrated with benzo(a)pyrene, which is shown in both abbreviated (on the right) and "long" forms (on the left). In the "long" form all carbon and hydrogen atoms are explicitly shown, with each carbon atom carrying the required four bonds. Note that the carbon atoms are arranged in rings of six each – they are "cycles" of carbon atoms, all of which contain six electrons (represented by the circle which was introduced earlier in connection with the discussion of benzene). Because PAHs contain several of these rings of carbon fused together, they are called "polycyclic."

Many PAHs are present in smoke and other products of combustion, and in pitches and tars. Some are carcinogens, others are not; carcinogenicity depends strongly upon details of chemical structure, specifically the ways in which the carbon rings are attached to each other. The PAHs are an important class of environmental pollutants, because of their widespread occurrence.

Innovations in chemical synthesis of dyes gave rise to one of the first major chemical industries. Following up on the work of the German physician Ludwig Rehn, who reported large "clusters" of bladder cancer cases among dye workers in the 1890s, occupational physicians began during the 1930s to study systematically the persisting high rate of this disease among dye workers. A decade or more of research by epidemiologists, occupational physicians, and chemists led to the identification of a number of substances called *aromatic amines* and *amino-azo* compounds as the culprits. The work of people such as Wilhelm Hueper on bladder cancers in the dye industry provided a major impetus to research and testing to identify other chemical carcinogens to which workers and the general public might become exposed. In 1937, Hueper and his associates at the National Cancer Institute (NCI) reported the experimental production of bladder tumors in dogs, from administration of the aromatic amine called 2-naphthylamine (see structure).

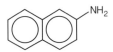

2-Naphthylamine

Hueper and several colleagues at the NCI were instrumental in drawing public attention to the issue of carcinogens in the workplace and the general environment during the two decades following the work on bladder cancer. Hueper's work and opinions were favorably cited many times by Rachel Carson in *Silent Spring*. Ms. Carson wrote:

Among the most eminent men in cancer research are many others who share Dr. Hueper's belief that malignant diseases can be reduced significantly by determined efforts to identify the environmental causes and to eliminate them or reduce their impact.

Carson's chapter entitled "One in Every Four" (referring to what was at the time the lifetime risk of cancer development in the US population), relies heavily on Hueper's work; she quotes him with great respect, and it is clear that most of her views on the subject are derived from him. It would be a mistake to conclude that Hueper alone was responsible for moving into full public view what science understood

at mid-century about environmental carcinogens. Scientists throughout the world began sounding alarms, though generally far quieter
ones, concerning the growing body of evidence that industrial products
and by-products, and even some natural chemicals, could be significant
contributors to human cancers. Rachel Carson's book, following by
two years the great cranberry scare mentioned in the Preface, heightened public interest and moved Congress to enact by the early 1970s
several major environmental laws that required regulatory controls
to be placed on many of the products that had been incriminated by
the work of Hueper and others. All of these laws called for stringent
controls on exposure to carcinogens.

On October 24, 1969, the Department of Health, Education, and
Welfare (precursor to Health and Human Services) established an
expert committee to advise the Department on carcinogens. The "Ad
Hoc Committee on the Evaluation of Low Levels of Environmental
Chemical Carcinogens" issued its report to the Surgeon General of the
Public Health Service in April, 1970. The committee worked under
the direction of Umberto Saffiotti, a physician trained in occupational
medicine at the University of Bologna, who had become Associate
Scientific Director for Carcinogenesis at the NCI. The committee's
report and recommendations summarized thinking, at least that of
scientists within the NIH and their immediate advisors, on the scientific evidence that had accumulated since Pott's findings in the late
eighteenth century. The committee's twelve recommendations, it can
be readily observed, also reflect much of what can be found in the
writings of Wilhelm Hueper, although they also derive from a number
of more limited expert panel reports commissioned by various arms
of the federal government since the early 1960s. The recommendations restate, with great emphasis, the special dangers of low level
carcinogen exposure. One recommendation states that "the principle
of zero tolerance . . . should be retained. . . ." The committee called for
the use of animal tests to identify carcinogens, emphasized the need
for testing at high doses, and also made some policy recommendations regarding the need for more comprehensive legislation to control
carcinogens. Perhaps the following passage from the report is the
defining one:

The effects of carcinogens on tissues appear irreversible. Exposure to small
doses of a carcinogen over a period of time results in a summation or poten-
tiation of effects. The fundamental characteristic which distinguishes the carcinogenic effect from other toxic effects is that the tissues affected do not seem

to return to their normal condition. This summation of effects in time and the long interval (latent period) which passes after tumor induction before the tumor becomes clinically manifest demonstrate that cancer can develop in man and in animals long after the causative agent has been in contact and disappeared.

It is, therefore, important to realize that the incidence of cancer in man today reflects exposure of 15 or more years ago; similarly, any increase of carcinogenic contaminants in man's environment today will reveal its carcinogenic effect some 15 or more years from now. For this reason it is urgent that every effort be made to detect and control sources of carcinogenic contamination of the environment well before damaging effects become evident in man. Similar concepts may apply to the need for evaluation of other chronic toxicity hazards. Environmental cancer remains one of the major disease problems of modern man.

## Cancer statistics

In the United States cancer caused 22.8% of all deaths in 2002, second to heart disease (28.5%) and slightly more than the next five causes (cerebrovascular disease, chronic respiratory disease, accidents, diabetes, and influenza and pneumonia) combined. Perhaps more significantly, death rates from cancer remained unchanged from 193.9 per 100 000 in 1950 to 193.4 per 100 000 in 2002, while death rates from heart disease declined from 586.8/100 000 to 240.1/100 000 during the same period. Death rates reflect both how much disease occurs and how well it is detected and treated. The relatively large decline in death rates from heart disease reflects large advances in prevention and treatment, which have, it is seen, not been nearly so successful in the case of cancer. Death rates vary by race, gender, and ethnicity. They are highest for African-American men, and lowest for white women.

Table 5.1 provides 2005 statistics from the American Cancer Society on new cases, not deaths, for the cancers that occur most frequently. Cancer incidence rates vary by race and gender in about the same way that death rates vary.

The incidence rates for various cancers can be derived from the data in Table 5.1 and other information, and expressed as lifetime probabilities, or risks. Thus, for example, these statistics tell us that, if incidence rates remain as they are now, then a male born today has a 50% risk of developing cancer over his lifetime, and a female has a

*Calculated Risks*

Table 5.1 *2005 Estimated New Cancer Cases^a (US)*

| Men | Percent of total cases | Women | Percent of total cases |
|---|---|---|---|
| 710 040 total cases | | 662 870 total cases | |
| Prostate | 33% | Breast | 32% |
| Lung and bronchus | 13% | Lung and bronchus | 12% |
| Colon and rectum | 10% | Colon and rectum | 11% |
| Urinary bladder | 7% | Uterine corpus | 6% |
| Melanoma of skin | 5% | Non-Hodgkin lymphoma | 4% |
| Non-Hodgkin lymphoma | 4% | Melanoma of skin | 4% |
| Kidney | 3% | Ovary | 3% |
| Leukemia | 3% | Thyroid | 3% |
| Oral cavity | 3% | Urinary bladder | 2% |
| Pancreas | 2% | Pancreas | 2% |
| All other sites | 17% | All other sites | 21% |

Source: American Cancer Society, 2005
^a Excludes basal and squamous cell skin cancers and in situ carcinomas except urinary bladder.

33% risk. Lifetime risks for men are largest for cancers of the prostate (1 in 6), lung and bronchus (1 in 13), colon and rectum (1 in 17) and urinary bladder (1 in 28). For women breast cancer leads the way (lifetime risk 1 in 7), followed by cancer of the lung and bronchus (1 in 18), colon and rectum (1 in 18), and uterus (1 in 38). These are the risks of developing cancers and say nothing about the chances of surviving them. Incidence rates such as these are critical to an understanding of how successful we are at preventing cancer. Death rates tell us more about the success of medical intervention.

## Causes of human cancer

With some exceptions, cancer experts generally cannot determine with high confidence the specific cause of cancer in an individual. At best they can understand the factors that contribute to the cancer rates observed in large populations. Differences in the rates of certain types of cancers in different regions of a country, different countries of the world, and in the same population studied at different times, provide some indication of the relative importance of various factors. Epidemiologists also learn a great deal from studies of specific exposure situations. Several trends emerge from these types of investigation:

(1) Somewhere between 70% and 90% of human cancers appear to be of environmental origin. Here "environmental" is used very broadly, and refers to anything not genetic. It refers not only to industrial chemicals and pollutants, but includes factors such as diet, sexual habits, smoking behavior, and natural and manmade radiation.

(2) Most cancers are not caused by individual carcinogenic factors, but by several factors. This view is consistent with our understanding that the gradual transformation of a normal cell to a malignant one occurs in steps, and that different agents may be involved at different steps (see the section on Mechanisms).

(3) In many cases a single factor may be so important that it is considered "the cause." Cigarette smoking, for example, is an important cause of lung cancer because in the absence of this habit about 85% of lung cancers would be avoided.

(4) It has become customary among cancer epidemiologists to talk about certain "lifestyle" factors as important contributors to cancer risk. Lifestyle factors (smoking, dietary patterns, alcohol consumption) are assumed to be largely under the control of individuals. These are distinguishable from factors that are less directly in the control of individuals (occupation, medicines, consumer products), and those over which individuals have little or no control (food additives, pesticides, environmental pollutants). Just how much control individuals have over the various "lifestyle factors" is of course much debated.

In 1981, two eminent British cancer experts, Sir Richard Doll and Richard Peto published a paper in the *Journal of the National Cancer Institute* entitled "The causes of cancer: Quantitative estimates of avoidable risks of cancer in the United States today." The authors drew upon a vast body of literature of the type mentioned above, and attempted to allocate the deaths caused by cancers among various responsible factors. The authors concluded that a certain percentage of human cancer deaths could be avoided if exposure to the responsible factors could be eliminated or controlled in some way, although the appropriate degree and nature of control for some of the "lifestyle" factors, especially diet, is still highly uncertain. The Doll and Peto estimates are presented in Table 5.2. The factors are listed in a somewhat different order from how they were listed by the original authors, because of our interest in clearly separating "lifestyle factors" (the first

Table 5.2 *Proportion of avoidable human cancer deaths for both sexes of the United States population*

R. Doll and R. Peto, 1981. *Journal of the National Cancer Institute*

| Factor | Percent of total cancer deaths | |
| --- | --- | --- |
| | Best estimate | Range |
| Tobacco | 30 | 25–40 |
| Alcohol | 3 | 2–4 |
| Diet | 35 | 10–70 |
| Reproductive and sexual behavior | 7 | 1–13 |
| Occupation | 4 | 2–8 |
| Food additives | less than 1 | minus 5–2 |
| Pollution | 2 | less than 1–5 |
| Industrial products | less than 1 | less than 1–2 |
| Sunlight, UV-light, other radiation | 3 | 2–4 |
| Medicines, medical procedures | 1 | 0.5–3 |
| TOTAL | 85–87 | |

*Note:* The remaining 13–15% are due to infectious agents (certain viruses and parasites) and some genetic factors that predispose certain individuals. The authors have recently reduced the percent associated with occupation to about 1. The "minus" end of the range for food additives takes into account the fact that some of these substances, particularly the antioxidants, may protect against certain cancers.

four listed) from those that are the more direct subject of this book; this change in no way distorts the original authors' conclusions.

What is striking about the Doll–Peto estimates is the relatively small fraction, perhaps 5–8% of human cancer deaths in the United States, that are attributable to industrial chemicals present in the work place, food, medicines, and involved in environmental pollution! If these estimates are correct, then no more than about 28 000–45 000 out of the nearly 560 000 annual cancer deaths in the United States are principally the result of the industrial products that are the main subjects of regulatory interest. While this number is hardly trivial, it pales in comparison with the more than 500 000 annual cancer deaths and nearly 1.4 million new cases diagnosed each year that might be avoided if we could get people to stop smoking, drinking to excess, and to eat the right diet (although, we are not sure what dietary regimen is optimum to avoid these cancers and whether such a regimen might not put people at increased risk from other diseases – dietary issues are extremely complex).

So why so much attention to what most people think of when they think of carcinogens – i.e., environmental chemicals of industrial origin? First, it ought to be made clear that while the Doll–Peto estimates are acknowledged by many cancer experts as close to the mark, and are in rough agreement with estimates made by others, they are still uncertain and have been criticized by some experts as possibly misleading. Other experts argue, for example, that since cancer can take several decades to develop, the full effect of the massive increase in industrial chemical production, usage, and waste disposal that occurred following World War II is not reflected in the cancer rate statistics relied upon by Doll and Peto, which were collected primarily in the 1970s. Also, Doll–Peto were only able to develop death rate data; incidence rates might have told a somewhat different story.

A 1990 publication in the British journal *The Lancet*, jointly authored by scientists from the United States, England, and the World Health Organization, contains an evaluation of more recent cancer mortality trends in six highly industrialized countries: France, West Germany, Italy, Japan, England and Wales, and the United States. These authors report a shift over the past two decades in certain patterns of cancer mortality. Specifically, they note that certain cancers – brain and other CNS cancers, breast cancer, multiple myeloma, kidney cancer, and non-Hodgkin lymphoma – have increased in both males and females, aged 55 and older, during this time at least in these countries. Stomach cancer continues to decline, as it has since the 1930s.

Incidence rates in children ages 0–14 have climbed from 11.5 per 100 000 to about 15 per 100 000 over the period 1975–2001; death rates during this period declined.

Epidemiologists attempt to use data on trends to generate hypotheses about cause and prevention, but that work is beyond what we are investigating in this book.

The Doll–Peto estimates, and others as well, nevertheless suggest a primary role for "lifestyle factors," not chemical pollution and industrial products, in cancer causation.

A second issue that apparently elevates public anxiety about industrial products to a high level concerns the fact that people feel they have little or no personal control over those products. Tobacco and alcohol usage, diet, reproductive and sexual habits, and sunlight exposure are to greater or lesser degrees within people's personal control, or at least they take personal responsibility for them, while they are involuntarily exposed to many industrial chemicals. People do not readily tolerate involuntarily imposed risks, even if these risks are small.

Of course, the U.S. Congress and legislatures all over the world have passed many laws requiring the regulatory control of exposure to industrial products, whether they be present as environmental pollutants, in the work place, or as ingredients in foods, medicines, and consumer products. Some of these laws single out carcinogens for special treatment, and thereby create heightened attention from regulators and the public.

So, the Doll–Peto estimates notwithstanding, it is necessary to continue to explore the scientific basis for concern about environmental chemicals that are carcinogenic. But as we move ahead on this topic we must keep in mind that we are dealing with only a piece of the total cancer problem, and are giving only cursory treatment to some perhaps overridingly important issues, such as the role of the diet and the issue of "multiple factors," most especially the genetic ones.

## Mechanisms of carcinogenesis

The groups of abnormally proliferating cells we call cancer can develop in any tissue or organ of the body. Most occur in *epithelial cells*, those that cover the surface of the body and that line all of the various organs; all such cancers are labeled *carcinomas*. More than 90% of cancers are of epithelial origin. About 2% of cancers are *sarcomas*; these arise in supporting structures of the body, such as fibrous tissues, skeletal muscle, bones and blood vessels. The remaining 8% are leukemias (malignancies of the hematopoietic system) and lymphomas (lymphatic system malignancies).

Neoplasms are, as we noted in Chapter 3, "new growths." But they are new in several highly destructive ways. Normal processes of cell replication within tissues occur in orderly, well-controlled ways. Neoplastic cells replicate wildly, without apparent controls. The relationships between the various types of cells within an organ are, if the organ is to function properly, also orderly; neoplastic cells, because of their disorderly replication patterns, can disrupt normal architecture and organ dysfunction can ensue.

It is also the case that the damage that occurs within cells to create neoplastic responses is heritable; it is found in all the progeny of affected cells. The damage, as we shall see, involves DNA, and the various cellular responses to this type of damage create "new growths." The proliferative responses leading to neoplasms are called *neoplastic transformations*. The term *tumor* is synonymous with neoplasm.

Neoplastic cells are described by pathologists as having varying *degrees of differentiation*: neoplasms that are "well differentiated" are those having characteristics close to those of normal cells, whereas poorly differentiated cells are radically different. *Benign tumors*, while abnormal, have a relatively high degree of differentiation, while malignancies are poorly differentiated. Other differences between benign and malignant tumors exist, the most important of which is the fact that malignancies can escape the tissues of their origin and migrate into blood vessels and lymphatics and thereby gain access to and colonize other organs – they *metastasize*.

## Multistage theory

It has seemed pretty clear for several decades, from both studies in humans and in experimental settings, that carcinogenesis is a *multi-stage* process. At the broadest level, the process can be thought of as one in which a normal cell is first converted to a permanently deranged cell, which is called a neoplastic cell, and a second sequence in which the neoplastic cell develops into a tumor, a neoplasm, that the pathologist can observe: neoplastic conversion and neoplastic development.

A 1954 publication in the *British Journal of Cancer* by Peter Armitage and Richard Doll[4] (whom we mentioned a few pages back) entitled: "The age distribution of cancer and a multi-stage theory of carcinogenesis" can be seen as a seminal event in the evolution of our understanding of the way in which cancers develop. Although there have been a number of successful modifications and refinements of the Armitage–Doll model in the 50 years since its publication, it is still seen as broadly correct.

The model was developed when little was known about the genetic and cellular mechanisms of carcinogenesis. The two professors drew upon earlier observations regarding the relationship between age and human cancer development and found clear patterns in observed, age-specific mortality rates that could be modeled mathematically based on a multi-step process (described below). That cancer might be initiated

---

[4] Sir Richard Doll passed away in July 2005 aged 92. He was an active research scientist to the very end.

from some type of mutation in the hereditary material of a *somatic* cell (those that do not play a role in the transfer of genetic information from one generation to the next) and that a series of events ("stages") following this mutation might lead to a neoplasm had been suggested by others, but in 1954 it was not the predominant view of cancer development.

For a fiftieth anniversary commemoration of the paper's publication, the *International Journal of Epidemiology* reprinted the original 1954 paper and invited commentary from several experts, including Sir Richard Doll. Doll wrote:

In 1948, when I began to work with Professor Bradford Hill at the Medical Research Council's Statistical Research Unit, ideas about the causes of cancer were still dominated by those of the great German pathologists of the 19<sup>th</sup> Century. . . . The idea that cancer might arise from a mutation in the hereditary material of a somatic cell had been suggested at least as early as 1930. . . . It was not, however, widely believed, which was surprising in view of the fact that Müller's demonstration, as long ago as 1927, that X-rays could produce hereditary mutations in fruit flies was universally applied and its application to humans was not questioned. . . .

The problem was that the mutational theory . . . postulated a single mutation and it was difficult to see how this could be made to account for some of the characteristic features of human cancer such as the rapid increase in incidence with age and the long latency period.

Doll then generously credits another British scientist, C. O. Nordling, for laying the groundwork, in a 1953 publication, for the move from single mutation theory to multistage theory. Sir Richard also points out that in their 1954 paper, Peter Armitage and he actually referred to "changes of state" and not mutations. He states that the two may have been thinking about mutations, but:

We did not want to describe the changes as such, however, as we did not want to put off the many cancer specialists, who were not happy with the mutational theory, from considering the idea that, whatever they were, the changes in a cell that made it the origin of a cancer clone were not a single event but a series of events, and that the factors that caused the changes to occur might vary in strength throughout an individual's life, irrespective of whether they were of external or internal origin.

He then explains, as we have already noted and will emphasize in the later chapters, not everybody exposed to carcinogens develops cancer; in fact even among heavily exposed populations only relatively small fractions of a population do so. This pattern is consistent with multistage theory (and its various modifications):

The fact that only, say, 20% of heavy cigarette smokers would develop cancer by 75 years of age . . . does not mean that 80% are genetically immune to the disease any more than the fact that usually only one cancer occurs in a given tissue implies that all the stem cells in the tissue that have not given rise to a malignant clone are also genetically immune. What it does mean is that whether an exposed subject does or does not develop cancer is largely a matter of luck; bad luck if the several necessary changes all occur in the same stem cell when there are several thousand such cells at risk, good luck if they don't. Personally I find that makes good sense, but many people apparently do not.

It is not entirely clear what Doll meant by the last phrase, but he may have been referring to an increasingly dominant view that some individuals are genetically predisposed to developing cancer when they become exposed to environmental causes. In any case, his brief, informal remarks which we have quoted at length provide an excellent synopsis of the history and basis for what we shall now discuss. We shall also be discussing methods used to estimate the "luck" (chance, or risk) Doll mentions.

How does the neoplastic cell come about? It seems pretty clear that the *initiating event* is brought about when the chemical carcinogen, which is in many, if not most, cases a metabolite and not the administered chemical, reaches a cell's nucleus and chemically reacts with DNA, the genetic material. This reaction constitutes DNA damage, an unwelcome event because this magnificent molecule controls the life of the cell and the integrity of its reproduction. Fortunately, cells have a tremendous capacity to repair DNA damage; these repair mechanisms have been at work probably since life began to evolve, because most types of cells are constantly being assaulted by DNA-damaging radiation and chemicals from many natural sources. If some of the damage is not repaired, and this happens because repair is not 100% efficient, and the cell undergoes replication when the damage is present, then the damage is passed on to the new cells, and can become permanent – a *mutation* has occurred. These mutations can take several different forms and result in different types of cellular alterations. Of particular interest these days is the role of so-called *oncogenes*. Oncogenes are the deranged twins of their normal counterparts, the proto-oncogenes. The latter are pieces of DNA that direct the synthesis of proteins necessary to the normal operation of cells. Sometimes, for reasons that are poorly understood, the functioning of the protein becomes uncontrolled. When this happens, the proto-oncogenes are said to have mutated to oncogenes and the cell in which this occurs begins its journey to a neoplastic transformation.

Experts refer to all cells that have been altered in their genetic features as *initiated* cells. The initiated cells might be fully neoplastic, in which case their proliferation takes place in a wild, uncontrolled fashion, or more commonly, they are only partially neoplastic. In the latter case the abnormal cells may still be under some control, and held in check by the actions of certain biological factors inherent in the organism in which the neoplastic conversion is occurring (the host). But some proliferation of these abnormal cells may also occur, perhaps brought about by continued chemical assault. Cell killing by chemically induced toxicity may cause tissues to produce more cells at a faster rate, i.e., to *proliferate*, and during this proliferation phase the abnormal cells can be further converted to fully neoplastic cells because of "genetic errors" that also proliferate – rapidly proliferating cells are at increased risk of this type of error.

In recent years, much interest has been generated by the discovery of genes that produce proteins having the capacity to keep dangerous cell proliferation under control. These are called *tumor suppressor genes*, and it seems they are more significant than proto-oncogenes in the development of human cancers. Damage to tumor suppressor genes, say through chemical interaction with a carcinogen, may result in rapid, uncontrolled cell proliferation of the kind that leads to malignancies. The most well-studied tumor suppressor gene is called *p53* and it plays a role in many human cancers, including those of the breast, the lungs, and the colon. The breast cancer susceptibility gene (BRAC) is another that generates suppressor proteins.

The protein produced under *p53*'s direction is, unfortunately, also referred to as *p53* (p for protein, 53 for its molecular weight in units of 1000). This suppressor protein has several critical functions in the life of a cell, among these a kind of "inspection" activity that prevents a cell from undergoing replication until any chromosome damage it has incurred has been repaired. If repair is adequate, p53 somehow signals the cell to reproduce. If damage is so extensive that adequate repair cannot be achieved, p53 signals the cell to commit suicide, an event called programmed cell death, or *apoptosis* (apo-ptosis, from the Greek word for leaves falling from a tree). It is easy to see why damage to the *p53* gene, especially in cells that had been subject to other types of genetic damage (had been initiated) could be devastating and could lead to extensive proliferation of mutated cells that should have been repaired or put out of commission. The study of factors involved in repair and in apoptosis are now major research endeavors in the search for cancer cause and cure.

Neoplastic cells may remain unobtrusive, under the regulatory control of various host factors. But when these controls, which may involve some intricate molecular communications between cells, break down, the abnormal cells can begin to grow and develop. The process is enhanced by chemicals called *promoters*, about which more later. After this *progression* to full-fledged malignancies takes place, usually by way of what pathologists refer to as the "benign" state. Promotion and progression are thus the two processes involved in creating a cancer out of a neoplastic cell. The ultimate neoplasm is thus a population of cells that arises from a single cell, what biologists refer to as a population of *clones*, which have expanded in numbers. The monoclonal origin of cancers is suggested by many studies of human and animal cancers.

This broad picture is considered pretty accurate by most cancer experts. Multiple stages are involved. They take place at different rates, and whether and at what rate they occur depends upon many factors, including at least:

(1) The concentration over time of the initiating carcinogenic chemical (typically a metabolite) at the cellular target.
(2) The presence of chemicals, which might be the carcinogen itself, its metabolites or even some other chemical – including some normal dietary components – that may restrict or enhance conversion or development, at several different points of the process.
(3) The influence of host factors, including cellular genetics and factors such as the host's hormonal and immune systems, that may either restrict or enhance neoplastic conversion or development.

What emerges here is a picture in which the carcinogenic process is influenced by a fairly long list of factors, and is either aided or inhibited by these factors. When we administer a chemical to lab animals and count tumors at the end of their lives, we are observing only two points, and not particularly interesting points, connected by a long sequence of molecular and cellular events.

We also might be accused of creating a highly artificial situation, because the cells of genetically homogeneous animals held under strict laboratory controls do not experience nearly the number and type of host and environmental influences experienced by people exposed to the same chemical. Perhaps the most that can be said about chemicals that are carcinogenic in laboratory settings is that they may, under some conditions, increase the risk of cancer in humans. Their relative importance in cancer development depends upon many other factors not accounted for in the laboratory experiment.

We might now examine some of the information that has contributed to this picture of the carcinogenic process.

## Electrophiles

This fancy word is used by chemists to describe organic molecules that contain groups of atoms that are highly susceptible to reaction with other groups, called nucleophiles. Nucleophiles are abundant in the giant DNA molecules. Elizabeth and James Miller and their students at the McCardle Cancer Research Center, at the University of Wisconsin, have produced a series of important scientific papers showing that many chemical carcinogens are metabolized to compounds having powerful electrophilic properties. The Millers' work was seminal in establishing that metabolism was essential to the action of most carcinogens, and also helped reveal the specific chemical reactions that take place on the DNA molecule.

Metabolism to electrophilic agents is highly important, for example, in the case of the aromatic amines referred to earlier. These amines are nucleophilic, not electrophilic in nature, and have little potential to react with DNA. But they can undergo metabolism to so-called N-hydroxy derivates. The amine group ($-NH_2$, where N is nitrogen) is converted to the N-hydroxy group ($-N-OH$) in cells. The presence of the $-OH$ changes the chemical character of the amine nitrogen atom and, through a sequence of events, the $-OH$ group departs the molecule, taking the electrons that comprise the $N-O$ bond with it, and leaving behind an extremely reactive, and electron-poor nitrogen atom – a highly electrophilic (electron-seeking) group. The electrophilic group sops up electrons from some of the nucleophilic centers of DNA. This reaction between nucleophilic and electrophilic groups can create DNA damage.

The N-hydroxylation reaction does not occur in guinea pigs, and this species does not develop cancers in response to aromatic amine exposure. Rats and most other species do, because they possess the enzymes necessary to create N-hydroxy metabolites. Humans also carry these enzymes, so we would expect them, unfortunately, to respond more like a rat than a guinea pig. This type of information explains a great deal.

Some examples will be seen in later sections of carcinogens that seem not to require metabolic activation to electrophiles, and which do not possess significant electrophilic properties themselves; these agents appear not to be involved in the initial DNA-damaging event, but rather at later stages of the neoplastic process.

## Genotoxicity

Toxicity to the gene is a seminal event in carcinogenesis, if the damage carries through to offspring of the cell that is assaulted initially. A serious omission thus far in this discussion of toxicity concerns the health implications of genotoxic events, of which initiation of carcinogenesis is but one possibility. This omission needs to be corrected before elaborating further on the role of gene toxicity in carcinogenesis.

First, we've talked so far about chemical changes in DNA, and this can certainly be an important form of damage. Chemicals that cause such changes, many of which are electrophiles, are called *mutagens*, and, as we have said, many mutagens are carcinogens. But DNA can be changed in other ways. Its very physical structure can be deranged by certain agents called *clastogens*. In some cases whole chromosomes (the molecular scaffolding of the cell's nucleus that carries the DNA molecules) can be added or lost, a condition called *aneuploidy*. Aneuploidy is often produced by chemicals that interfere with the mechanics of cell division. Not only can genetic damage increase the risk of cancer development, but can be deleterious in other ways. Cell death or abnormalities may occur because some of the protein molecules essential to their existence, and which are created under the direction of DNA, can no longer be produced with fidelity to the cell's original blueprint.

If the genetic damage occurs in germ (as against somatic) cells, reproductive failure may occur, or if it does not, the abnormal cells are carried forward and may create abnormal offspring. A permanent genetic abnormality may continue in one generation after another; this is a true heritable mutation. Whether environmental chemicals contribute significantly to this type of inherited mutational change is not clear at the present time. Assessing human mutagenic risk for chemicals has been little explored, much less so than mutagenic risks incurred by radiation. The latter have been under study since

H. J. Müller's discovery in 1927 that radiation could induce mutations in living organisms. Whether congenital diseases such as Down's syndrome (in which an extra copy of a particular chromosome, or aneuploidy, is present) are much influenced by chemical mutagens in the environment remains an active area of study by genetic toxicologists.

One of the more interesting pastimes of the genetic toxicologist is the development of simple tests to detect the capacity of a chemical to produce genetic damage. The challenge is to find the quickest, simplest procedure that is also telling. Many of these tests are performed in glassware (in vitro), outside the whole animal. Microorganisms or cells from animals and even from people are placed in liquids containing the nutrients necessary for their growth, and suspect chemicals are added to the liquid. The genetic toxicologists have found a range of clever ways to detect genetic damage in cells grown in this way. They have even found ways not only to test chemicals, but also metabolites of those chemicals; in effect, a means is found to incorporate those enzymes responsible for the mammalian metabolism of the chemicals into the in vitro system. Many of these tests can be performed in a matter of hours or a few days at relatively small cost.

In vivo mutagenicity tests are also plentiful, but because they involve whole animals, they are generally more time-consuming (days to several weeks) and expensive.

Professor Bruce Ames, a biochemist at the University of California at Berkeley is one of the pioneers of this type of short-term testing. The Ames Test, as it is called, is now widely used, typically as one of several short-term tests that constitute a series of tests, or battery. A battery is thought necessary because no single test is adequate to detect all types of genotoxicity. The Ames Test involves the use of mutant strains of a common bacterium, *Salmonella typhimurium*, that "back-mutate" to their normal state in the presence of a mutagenic chemical or metabolite. Many other bacterial and mammalian cell systems have been made available for this type of testing.

Not too long ago cancer specialists were excited about the prospect of using some of these short-term tests to detect carcinogens. They could replace the very expensive and time-consuming animal bioassay reviewed in the next chapter. After all, these tests detected genotoxicity, an initiating event in carcinogenesis. Genotoxic agents ought to be carcinogens, and those with no genotoxic activity should not be.

Well, this was too simple. Perhaps a high proportion of genotoxic agents are carcinogens, but toxicologists have learned that many chemicals having little or no gene-damaging power are also carcinogenic. Using genotoxicity as the sole criterion for detecting carcinogens would result in missing a number of possibly important agents (although, as shall be seen in later chapters, it may be that the genotoxic carcinogens are riskier at very low doses than those that act at later stages of the carcinogenic process and that are not genotoxic). Carcinogens that act not as initiators of carcinogenicity, but at later stages of the process, apparently do so through mechanisms not involving gene damage. Some experts have introduced the categories of genotoxic and epigenetic ("epi" meaning "outside of") carcinogens. "Initiators" and "promoters" of the process might be other terms for these categories. These general categories are widely acknowledged, although most recognize that a dual categorization may be too simple, and may obscure some important distinctions. Still, the general notion that some carcinogens act at "early stages" and others at "later stages," and that some carcinogens may in fact act at both early and later stages, is highly important and certainly fits well within the multistage model. In fact, the model was constructed in part upon experimental observations that carcinogens could indeed act in different ways, at different steps of the process. Not all carcinogens are the same, not by any means.

## Promotion

Peyton Rous in his laboratories at the Rockefeller University, and Isaac Berenblum of the Weitzman Institute in Israel, together with Philipe Shubik were, back in the 1940s, the first to reveal that certain chemicals, apparently not carcinogenic themselves, could somehow greatly enhance the effects of other substances known to be carcinogenic. The classic experiment involves application of a highly carcinogenic polycyclic aromatic hydrocarbon (PAH) to the skin of shaved mice, followed by application of substances called phorbol esters. Phorbol esters are complex organic compounds found in croton oil, a natural extract from the seeds of the croton plant. The PAH can, when applied in sufficient amounts, produce excess skin tumors in great abundance. This might be called local carcinogenicity – the mouse skin bioassay has great utility for some types of experimental cancer work because it is relatively rapid and the development of neoplasms is easy to monitor, although not many carcinogens produce skin tumors when applied in this fashion. When the PAH dose is dropped to a low level, such that no, or only a few, skin tumors would be expected, and the phorbol esters are later applied to the area of the skin treated with the PAH, the yield of tumors climbs dramatically. The phorbol esters are themselves carcinogenically inactive on the mouse skin, but *promote* the development of cells *initiated* by the PAH metabolites, which are highly genotoxic. These findings were important to the Armitage–Doll formulation of multistage theory.

Now most people are not exposed to PAHs in this way, and no one is exposed to the phorbol esters, which are strictly laboratory chemicals. But this is not the point. The Berenblum–Shubik promotion studies, and hundreds more that have been explored in the past five decades, have contributed substantially to our understanding of the carcinogenic process. Some chemicals appear to possess both initiating and promoting properties – they are "complete" carcinogens – while others seem to be primarily involved in the promotion stage of the process.

Although the molecular and cellular events associated with initiation are, as we have indicated, quite well understood, promotion is still fairly mysterious. Somehow the presence of the promoter leads to a breakdown in the system of controls that tissues use to restrict the growth of cells that have undergone neoplastic conversion. Molecular biologists have been examining the processes by which cells communicate and interact with each other – their means for regulating one another's behavior. These interactions occur through some portions of the cells' membrane, known as gap junctions, and it appears that some promoters act at the gap junctions to interrupt the intercellular communication essential to controlling the behavior of aberrant cells.

Promoters may be prominent players in several of the most important human cancers. It appears that bile acids, which are the major components of bile, are promoters for cancer of the large intestine. High fat diets greatly increase bile acid flow through the colon. High fiber intake helps eliminate bile acids, and thus reduce the risk of large bowel cancers. The experimental demonstration that bile acids are promoters of colon cancer is consistent with epidemiological observations that increasing fat intake increases the risk of colon cancer, although establishing a causal link between fat intake and colon cancer continues to be elusive.

Promoters may have a significant bearing on several other human cancers, including those of the breast, ovary, and prostate. It is also of more than a little interest that many components of the diet, in addition to fat and fiber, can significantly modify the response of experimental animals to carcinogen exposure, in some cases enhancing and in others inhibiting the response. In fact, simply restricting total caloric intake can reduce tumor yield, especially in those tissues such as the breast, ovary, and endometrium that are under the control of sex hormones. Some of these dietary influences are no doubt promotional in nature, but most are not well understood and it remains difficult to acquire a clear picture from the epidemiological evidence regarding diet and cancer. It is quite apparent, however, that nutrient and non-nutrient components of the diet have a major influence on cancer rates, as suggested by the Doll–Peto estimates (Table 5.2), and the experimental work on diet–carcinogen interactions is beginning to reveal why this is so.

Substances such as promoters that interfere with cell-to-cell communication allow cancer cells to proliferate wildly. But cell proliferation can be induced by other means as well. Toxicity or other types of injury to tissues can result in a proliferative response. So can certain natural and synthetic hormones, such as estrogens, cause proliferation of certain tissues, such as the breast. Chronic viral infections may cause cell killing and its consequence is cell proliferation. It appears that sustained chronic proliferation induced in any of these ways, either by agents foreign to the body or some, such as the estrogens, that are natural to it, can increase tumor growth.

In some cancer bioassays, as we shall see, the highest dose tested is referred to as the maximum tolerated dose (MTD). In some cases the MTD may be sufficiently high to cause toxicity and cell proliferation, putting affected tissues at extra risk of cancer. It is also the case that rapidly proliferating cells, even if they have not been initiated, are at increased risk of conversion to neoplastic cells. They are more prone to the mutational events that are always present naturally. Professor Bruce Ames of the University of California at Berkeley holds that "a high percentage of all chemicals, both man-made and natural, will cause cell proliferation at the MTD and increase tumor incidence." This is a refined way of saying that almost everything will cause cancer if the dose is pushed high enough, to a level sufficient to cause extensive and sustained cell proliferation. Whether this is true remains a subject of intense debate. There are some cases in which this mechanism of cancer development has been taken into account in the risk assessment process, and we shall encounter one in Chapter 9.

## Some implications

Some important conclusions emerge even from this rudimentary profile of mechanisms. Pharmacokinetics and metabolism are as significant in carcinogenesis as they are in the production of other forms of toxicity, so a thorough evaluation of risk would require knowledge of species' differences in ADME, and the influence of the size of the dose on pharmacokinetic behavior.

Initiating events involve gene damage, and this may result in fixation in the cell's genetic material of a permanent abnormality. This feature of carcinogenesis perhaps makes it different in kind from most other forms of toxicity. Here the chemical insult occurs and the damage it produces may remain in cells even if exposure to the insulting chemical ceases. If doses of the genotoxic agent keep piling up, so do the numbers of those permanent changes. This rather frightening picture is made less so when we recall that cells have a tremendous capacity to repair DNA damage before it becomes fixed, so that not every damaging event, in fact perhaps only a tiny fraction of them, actually translates to a mutation, and only a small fraction of mutations will likely occur at sites that are critical for the development of cancer.

One implication of this view of initiation – and an exceedingly important one – is expressed in the "no-threshold" hypothesis for carcinogens. Any amount of a DNA-damaging chemical that reaches its target (the DNA) can increase the probability of converting a cell to a neoplastic state. This does not mean that every such event will *cause* a neoplastic conversion, but only that the *probability*, or *risk*, of that occurrence becomes greater than zero as soon as the effective target-site concentration of the gene-damaging chemical is reached, and that the risk increases with increasing target-site concentration. Sir Richard Doll's comments about the role of luck are right on target.

Actually, the notion that human cancers might result from exceedingly small doses arose first in connection with radiation-induced malignancies. In the 1950s E. B. Lewis of the California Institute of Technology proposed, based on studies of leukemia rates among Japanese atomic bomb survivors and cancer rates among radiologists, that cancer risks might exist at all doses greater than zero, and that a linear dose–response relation is to be expected.

A linear, no-threshold model might be appropriate to describe the dose–response relation for a genotoxic carcinogen, but it is less clear that it is suitable for promoters that act through non-genotoxic mechanisms. Some experts contend that sustained, high level dosing is needed to promote carcinogenesis – the dose needs to be sufficiently large to induce a persistent state of cell proliferation or a breakdown in cell-to-cell communication. Until a threshold dose for these toxic effects is exceeded, these experts suggest, significant enhancement of the carcinogenic process is unexpected. All this gets fuzzy when we consider that some complete carcinogens possess both initiating and promoting properties. A fuller discussion of these and selected dose–response issues is reserved for the chapters on risk assessment.

Mechanisms of toxic actions, as we have already seen, are not all of the type that seems to hold for carcinogens. Some toxic agents act by interfering with the cell's capacity to generate and use energy; its basic metabolic arrangements can be disrupted, leaving it either to die or to operate improperly. Promoters of carcinogenesis are not the only type of toxic agent that can interfere with cell-to-cell communication, and thereby impair the health of a tissue. The cell membrane can be damaged in several different ways by certain toxic chemicals, and this can touch off a series of deleterious events. These various mechanisms, unlike those associated with the production of mutations, all seem to require that some minimum dose of the toxic agent or its metabolite reach the cellular or tissue target, in many cases for extended periods of time – the threshold dose for toxicity needs to be exceeded. Toxicologists are by no means certain of this, because mechanisms are not yet worked out in sufficient detail. But, this view guides most current risk assessments.

The study of toxicity mechanisms will continue. Scientists from the basic disciplines of molecular biology, genetics, and biochemistry are becoming increasingly involved in the field of toxicology, and this is a highly desirable trend. The practical payoff, which is to translate what the "molecular toxicologists" are learning about mechanisms into more accurate characterizations of human risk, is not quite around the corner, but it is surely somewhere in the next block.

Against this background we now turn to the problem of identifying which specific chemicals can bring about the production of neoplasms.

# 6

# Identifying carcinogens

Beginning in the late 1940s, when Dr. Hueper established an Environmental Cancer Section at the NCI, and continuing to this day, a major program of carcinogen identification using animal tests has been conducted at the National Institutes of Health (NIH). At the present time most of this testing is conducted under the auspices of a government-wide activity called the National Toxicology Program (NTP) centered at another NIH unit, the National Institutes of Environmental Health Sciences (NIEHS) in Research Triangle Park, North Carolina. Several hundred chemicals have been tested for carcinogenicity by the NTP. The NTP and other federal health and regulatory agencies also sponsor epidemiology studies and other investigations into the underlying chemical and biological mechanisms by which some chemicals transform normal cells into malignant ones. In addition to government-supported work, there is substantial industry-sponsored testing and research of the same type, some of it performed because of regulatory requirements. The field of chemical carcinogenesis is a vast scientific enterprise, not only in the United States, but throughout the world.

One important part of this vast enterprise is the International Agency for Research on Cancer (IARC), a part of the World Health Organization, headquartered in Lyon, France. One of IARC's many activities involves convening meetings of scientific experts from across the world to examine published scientific work relating to the carcinogenicity of various chemicals. The IARC periodically publishes the results from the deliberations of these working groups. The agency

Table 6.1 *Some of the 95 chemicals and occupational exposures currently listed by IARC as carcinogenic to humans*

Note that in many cases data on cancer rates were collected under exposure conditions that no longer exist.

Some occupational exposures
    Boot and shoe manufacture (certain exposures)
    Furniture manufacture (wood dusts)
    Nickel refining
    Rubber industry (certain occupations)
    Underground hematite mining, when radon exposure exists

Some chemicals
    Aflatoxins
    Arsenic and arsenic compounds
    Asbestos (when inhaled)
    Chromium [VI] compounds (when inhaled)
    Benzene
    Diethylstilbestrol (DES)
    2-Naphthylamine, benzidine (starting materials for manufacture of
        certain dyes)
    Vinyl chloride (starting material for PVC plastic manufacture)
    Mustard gas

Some chemical mixtures
    Tobacco smoke
    Smokeless tobacco products
    Soots, tars, mineral oils[a]
    Wood dust

[a] Mineral oils now in commercial production generally do not have the PAH content they had at the time the evidence of carcinogenicity was gathered.

also categorizes chemicals based on the nature and extent of available scientific evidence concerning their carcinogenic activity. Evidence is labelled as "sufficient," "limited," or "inadequate," and the reviewed chemicals are grouped into these categories; distinctions are made between evidence based on epidemiology studies and that based on studies in laboratory animals.

In Table 6.1 are listed some of the chemicals and occupational settings IARC has categorized as carcinogenic to humans – the data from all the available epidemiology studies were considered sufficient by the expert groups to conclude that a *causal* relationship exists between exposure to the chemical or occupational setting and some form of human cancer.

Several comments should be made on the contents of Table 6.1. First, it is not complete; IARC now lists a total of 95 chemicals, chemical mixtures, and chemical processes as carcinogenic to humans; the agency lists another several dozen or so as having limited evidence of human carcinogenicity. Second, occupational exposures are listed instead of individual chemicals for those cases in which several chemicals may be involved in the exposure, and it is not clear which are responsible for the cancer excesses observed in the workers studied. Third, the listed chemicals and occupational exposures have been demonstrated to be human carcinogens only for the specific groups of individuals upon whom observations were made by epidemiologists; whether they increase cancer risk for other individuals exposed under quite different conditions is a separate question, to be treated later in this chapter and in the chapters on risk assessment. Finally, it should be emphasized that there are many additional animal carcinogens listed by IARC that may also pose a carcinogenic risk to humans, but for which sufficient epidemiology data have not been collected; keep in mind that a chemical can be demonstrated to be a human carcinogen only if an opportunity exists to study it in exposed humans in a systematic way, and such opportunities are not always available.

How evidence becomes "sufficient" to establish causation is a major topic for discussion in the rest of this chapter, as is the relationship between animal findings and human risk.

The accumulating evidence that carcinogenesis is a multistage process, and that different factors or substances may affect the several transitions a cell has to undergo to arrive at a malignant state, would seem to make obsolete any simple ideas about a "cause" of cancer. So, what does it mean, exactly, when the WHO's International Agency for Research on Cancer lists smoking as a cause of lung cancer, or the EPA lists benzene as a cause of leukemia? To tackle these questions, and prepare for the coming journey through the world of risk assessment, we need to provide a more systematic and critical look at the two areas of science – cancer epidemiology and experimental carcinogenesis – from which emerges the evidence we use to label various substances as carcinogens.

## Causation and prevention – how does epidemiology help?

Here is a definition of epidemiology that is as good as I have found:

Epidemiology is the study of the distribution and determinants of health-related states or events in specified populations, and the application of this study to the control of health problems.

First, note that epidemiology is by no means limited to the study of cancer, but is used to understand many types of infectious and non-infectious diseases. It is also, as is clear from this definition, not limited to unhealthy states, but includes the study of the determinants of good health. It is also concerned with *populations*, not individuals, and so does not involve the same methods and approaches used by physicians in diagnosing medical conditions, and even their possible causes, in individual patients. It is focused on populations, and results from epidemiology studies pertain only to populations. So, if an epidemiologist claims that smoking two packs of cigarettes each day for forty years causes a tenfold increase in the chance of developing lung cancer by age 55, he or she is referring to an increased risk in a population of smokers, and is saying little about any specific individual in that population.

What good is such information? Well, it is critical, essential even, if we wish to improve the public health by preventing disease. The last phrase of the definition makes this point, and it is not an empty tag line. If epidemiologists can determine which populations are most affected by certain disease states, and then do some detective work to figure out how those populations differ from others that suffer less disease, then they have a chance of identifying the conditions that contribute to that disease state. Once epidemiologists have made this determination, public health experts or regulators can begin to work on ways to reduce the presence of the responsible conditions, be they certain chemical exposures, infectious agents, lifestyle factors, or physical phenomena such as radiation. Epidemiology is the key to population-based preventive medicine.

Of course individual people can, for causes under their personal control, also benefit from the results of epidemiology studies; even though epidemiology results apply to populations, individuals at least have an increased *chance* of benefiting by avoiding the substances or factors identified as causes of disease, especially when, as in the case of smoking, the population risk from the causative agent is unusually large. Individuals would be well advised to avoid smoking – they will hugely increase their odds of avoiding the many smoking-related diseases if they do so. As a generalization, using epidemiology results to fashion personal behavior is most useful, most likely to pay off, for causes the epidemiologists have identified as greatly increasing risks – smoking, excessive alcohol consumption, diets high in saturated fat, driving without a seat belt, and a few others. As epidemiologists begin to report on small and even mid-sized risks, the certainty of their findings declines and the benefits of individual action become less predictable.

It is no secret that epidemiologists have a difficult time coming up with conclusive results – everyone is familiar with the "health advice-of-the-month" syndrome, wherein what epidemiologists said was good for you last month is, according to the latest study, now proven to cause some debilitating disease. Although epidemiologists and those who try to report epidemiology results are often ineffective at describing for the public exactly what various studies tell us, it is through a look at the nature of the science itself, and the limits against which it is working, that the fundamental reasons for confusion, apparent or real contradiction, and controversy are to be understood. As elements of the science and its methods of study are set out in the following, its limits will become evident. We would like to know for sure whether certain substances to which we may be exposed can cause disease, especially if the disease is cancer, but the only means we have available to look at causes, epidemiological studies, cannot usually get us to a "for sure" answer without a long and difficult scientific struggle.

## Describing the world of cancer

In the beginning, before there is analysis, there must be accurate description. How much cancer is there, and how do rates of occurrence vary geographically, and between sexes, and with age? How do rates of different types of cancer vary over time, and what happens to the rates that occur in specific groups of people when they move from one geographic location to another? Information describing these types of differences and trends – which can be compiled with accuracy only when cancer registry information is reliable – are enormously beneficial in providing clues to the causes of cancer. The statistical data presented in Chapter 5 arose from these types of studies.

Here it should be emphasized that two types of cancer registry information are available, and afford somewhat different kinds of understanding. The easiest information to compile is that pertaining to death rates, and its accuracy depends upon reliable medical diagnoses of cause of death and the careful recording of that information. Cancer death rates are, however, not the best types of data for getting to the issue of causation. Death rates reflect both incidence of the disease (how much cancer occurs) and the results of medical interventions. Incidence data are not influenced by the effects of treatment and are more useful if the goal of the epidemiologist is to gain some understanding of cause.

It is easy to imagine that incidence data – how many new cases of specific types of cancer occur in a given population over some specified period of time – are more difficult to come by, and it is only in recent years that reliable incidence data have been available. In many countries they are still not readily available. In any case, epidemiologists need to worry about differences in incidence if they are to uncover underlying causes.

Descriptive epidemiology is, however, only the beginning, because it cannot provide significant information regarding specific causes or determinants of cancer. It has been ascertained, for example, through a series of careful descriptive studies, that, after a couple of generations, Japanese immigrants to California exhibited changes in their rates of stomach cancer (declines) and breast cancer (increases) that brought them closer to those found in native Californians. This type of information is profoundly important in establishing that environment is a highly significant influence on cancer risk. It reveals little, however, about the possible underlying causes of these changes, and therefore little of direct public health benefit. Do we recommend, from the results of such descriptive studies, that Japanese who wish to reduce their risk of contracting stomach cancer should move to California? Obviously not. What should follow from such observations is more study of the specific environmental factors – certain dietary habits, perhaps – that result in the excess risk of stomach cancer that is seen in Japan.

There is a striking relationship between the levels of meat consumption in different countries and the rates of colon cancers in those countries. But does that observation alone establish that meat consumption is causative? Is this type of information, of itself, a sufficient basis for public health authorities to recommend that the citizens of high risk countries such as New Zealand and the United States, for example, reduce their meat consumption to something like the levels in Japan and Finland, both very low risk countries? The answer – this type of descriptive data is not difficult to acquire – is almost certainly no. It would be a mistake to interpret information of this type as establishing that meat consumption is a significant cause or contributor to colon cancer risk. There are several reasons why such a conclusion is fallacious.

First, there is no way to know whether, within each country, those groups of people who eat more meat exhibit greater rates of colon cancer than those who eat less or even no meat. Data concerning total or average meat consumption provide no clues about how that consumption is distributed among members of the population; moreover, the distribution patterns in different countries are surely variable. The data are also uninformative on the question of duration of meat consumption – they do not reveal whether rates rise with increasing duration of meat intake.

Perhaps meat consumption is related to some other factor and it is that other factor that is the real culprit. Do heavy meat eaters consume more alcohol than those who consume less? Do heavy meat eaters consume less fiber? Even if the latter were true, and we were to construct a figure relating fiber intake to colon cancer, and found that country rates went up as fiber intake declined, we would still fall short of demonstrating causal links between decreased fiber consumption and increased colon cancer rates.

These types of data are too often over-interpreted, and taken to provide more insight than is possible. The well-trained epidemiologist understands the limits, knows that, at most, such data only suggest where to look for more telling information. Descriptive studies are exceedingly important clues to the causation problem, but are no more than that.

Other clues arise from what are called case-reports. An English physician, John Hutchinson, reported, back in 1887, on unusual skin growths, including cancers, on individuals given Fowler's Solution (arsenic oxide) as a medicine. In 1970 Arthur Herbst of Harvard University identified seven young women who, over the course of one year, were diagnosed with a form of vaginal cancer that was extremely rare for their age group, and he suspected and reported that the use of the synthetic estrogen diethylstilbestrol (DES) by their mothers during pregnancy was possibly responsible. Unusual medical conditions such as these are often reported in medical journals, and the physicians reporting them may or may not have a guess about their cause. If the condition is not cancer or some other disease that typically requires a long period to develop, and the physician notes some common exposures (a particular medicine, say, or some occupational exposure) that are in the patients' recent history, his or her guess about the cause may be worth further study. Also, if the cancer is an extremely rare one and common exposures, even if they occurred many years in the past, are readily discernible, the physician's guess may be a reasonable one. If, however, the cancers are not extremely rare, the physician's guess about cause is likely to be a poor one, even if he or she thinks that all the patients experienced some common exposure, even an exposure that is unusual. Case-reports, or even reports of a series of cancer cases, even if they are rare cancers or if the cases experienced common and unusual exposures, are almost always insufficient to establish causation. At most, they can provide important clues for follow-up studies.

## From descriptive to analytical studies

The problem with the case-report approach is that there is no comparison group, and no control subjects. Cancers of different types occur at different rates within a population, and these rates also vary with age, geographic location, and often with sex, race, and ethnicity. Sometimes, in certain groups of people, rates are in excess of what is normally expected, because those people are exposed to certain factors that increase cancer risk. If such factors are to be identified, the epidemiologist must be able to demonstrate that cancer rates in the exposed population are, indeed, in excess of what is normally expected – a comparison group, unexposed to the factor, is the source of information on what is normal. A demonstration that a group of individuals experiencing excess cancer rates is exposed to some factor and that the control group is not so exposed is, for a number of reasons we shall get to, insufficient to establish that the factor is the cause, but such a demonstration is the starting point – the essential first step.

A demonstration of this nature is said to establish an *association* between the factor and the excess cancers. Two events – in this case the factor and the excess cancers – are said to be associated when they occur together more frequently than they should if *chance alone* were operating. We need to turn to statisticians to figure out whether the odds are greater than those associated with a chance occurrence. Association is not causation, but demonstrating that an association exists gets the epidemiologist part of the way to an answer.

Before describing the path that gets us the rest of the way to an answer, we shall describe the kinds of studies necessary to determine whether associations exist. Because the epidemiologist is attempting to understand effects in human populations, he or she cannot undertake experiments. Experiments are reserved for the laboratory, where it is possible to deliberately expose a group of animals (or cells in culture) to a suspect agent, and to compare the responses of animals in that group with those in a group of unexposed animals. In the lab setting, if the experiment is well performed, the exposed and unexposed animals are, in every other respect – genetic background, diet, environment – identical. So, in such circumstances, demonstration that the exposed group develops more neoplasms of a certain type than the unexposed group, not only establishes an association, but also provides strong evidence for causation. Obviously, for ethical reasons, we could never conduct such an experiment in human populations.

Note also that, even if some evil epidemiologists could undertake such an experiment in humans, there is virtually no way two human populations could be assembled and matched in the way lab animal populations can be matched, so that causation would remain difficult for the epidemiologist to establish. This "matching" problem begins to explain why there is much distance to travel even after an association is demonstrated.

One other aspect of epidemiological science should be mentioned before we examine the types of studies that can ethically be undertaken to get us on the road to a determination of causation factors in cancer. There is a branch of the science in which human "experiments" are, in fact, undertaken. Such experiments are known as clinical trials, and most are devoted to studying the potential benefits of drugs, other medical therapies, and certain dietary regimens. Such trials are entered only after it is established to the extent possible, through experiments in animals, that the drug or other therapy will do no harm to the study subjects. Often, unexpected "side effects" will be noted during such trials, and information on such effects is critical to understanding the drug's or therapy's "risk–benefit" profile, but such trials cannot be entered unless there is good reason to believe harm will not occur. We treat this subject more fully in Chapter 8.

Epidemiologists also conduct planned community trials to learn about the benefits of various nutritional programs or other efforts to improve public health. Again, such experiments with human populations are permitted only if there is a high degree of certainty that no-one will be harmed.

So what, exactly, can epidemiologists do to learn whether certain factors – chemicals or chemical mixtures, occupations, dietary patterns, lifestyle factors, infectious agents, radiation – cause cancer? Well, they can try to set up something approximating a controlled experiment in populations of individuals that, for various reasons, experience exposures to such factors at greater rates, or in greater amounts, or for longer periods of time, than do other populations of people. The epidemiologist is doing nothing to create the excess exposure, but is trying, through the use of some careful analytical methods, to determine whether excess exposure makes a difference with respect to cancer or other types of risks.

## Cohort and case–control studies

Epidemiologists can undertake two types of *analytical studies* (the term is used to distinguish them from descriptive studies and from case-report investigations). The first type involves individuals with common exposures who are followed over time to see whether they develop more cancers of specific types, during a specified period of time, than do unexposed individuals. This type is called a *cohort* study. The second type begins with cases of a specific type of cancer and proceeds to a determination of whether these individual cases have certain exposures in common, and whether these exposures are more intense, or of longer duration, than those of individuals who do not have cancer. This is termed a *case–control* study. *Both require control groups*. As noted (and this should be obvious) it is never possible to develop a complete matching of exposed and unexposed groups (in a cohort study) or of cases and controls (in a case–control study), but by exercising sufficient care, the known important differences can often be taken into account.

Both types of analytical studies can be thought of as natural experiments, where the epidemiologist is trying to extract knowledge from situations that have already occurred (*retrospective* analysis) or that are ongoing (*prospective* analysis).

Perhaps the first "modern" analytical studies were those reported in the 1950s on the relationship between smoking and lung cancer. Richard Doll (he appears again!) and Bradford Hill, of England, were the authors of cohort studies of smokers, and Ernst Wynder and Evarts Graham, of the United States, undertook and reported case–control studies of lung cancer cases. The Doll–Hill studies are, in their methodology and modes of data analysis, remarkably like cohort studies conducted today, and represent an extraordinary advance over previous attempts to establish causal links between environmental factors and cancer. The Wynder–Graham studies, while important in establishing links between smoking and lung cancer, were more like elaborate and extensive evaluations of a series of case reports, but still had many of the elements of modern case–control studies.

Cohort cancer studies are expensive, time-consuming, and are not particularly useful for detecting small or even moderate-sized risks, but epidemiologists generally place high confidence in results from them – well-done cohort studies are the "gold-standard" in non-clinical epidemiology. Case–control studies are less expensive and time consuming, are best-suited for studying rare or moderately rare cancers, but, as we shall see, results from them are often less clear-cut than those from cohort studies.

The first goal of the epidemiologist, whether engaged in a cohort or a case–control investigation, is to determine whether there are meaningful differences between the groups studied. In the case of the cohort study, the goal is to determine whether the exposed group experiences a greater risk of cancer than does the control group. Actually, each type of cancer is analyzed separately, because most exposures are expected to affect only one or, at most, a few types of cancer. There is no biological basis for supposing that some individual substance, or even a complex mixture such as cigarette smoke, will increase the risks of many or all types of cancer. Many cohort studies will reveal that overall cancer risks do not differ between exposed and unexposed groups, but that the risk for a particular type of cancer is increased (it will also be the case that risks for other types will sometimes be less in the exposed group). Cohort studies thus focus on how the risks of specific types of neoplasms are affected by exposure to the substance under investigation. For each type of cancer, the epidemiologist lists the *relative risk* (RR) – the rate of occurrence of cancer in the exposed group divided by the rate in the control group. A relative risk of 1.0 means there is no difference between the two groups in cancer risk; values greater than 1.0 mean there is an elevated risk for that type of cancer in the exposed group, and values less than 1.0 mean the opposite. The epidemiologist then consults his or her favorite statistician (who is probably him- or herself), because these relative risk numbers do not tell the whole story.

Recall that the epidemiologist is attempting to ascertain, in the case of the cohort studies, whether the two phenomena under study – the exposure to the suspect substance and the risk of cancer in the exposed group – are associated. That is, he or she is trying to find out whether these two phenomena occur together more frequently than they would if chance alone were involved. The simple observation that the relative risk is greater than 1.0 is insufficient to make this determination.

As it is in the case of animal toxicology experiments discussed in Chapter 3, the problem is one of "sample size," and this is one of the thorny problems statisticians are born to help us with. It is not much different from the problem of estimating the odds of various outcomes from flipping a fair coin. If it is fair, exactly half of the flips will come up heads and half tails. If, however, someone were to flip a coin only 10 times, it would not be a great surprise if heads and tails did not come up five times each. Even with 100 flips, a 50–50 split might not be seen; this would be more surprising than the result with 10 flips, but not greatly so. As the number of flips increases it would be expected that an exact 50–50 split would become more and more likely; if we flipped an infinite number of times we would no doubt arrive at an even split (each half of which, according to mathematicians who fantasize over things like the infinite, would also be an infinite number).

The point of all this is that we would not reasonably assume a coin to be an unfair one if, in a relatively small number of flips, we did not observe a 50–50 split. Rather, we would do a little statistical work, and figure out, for a given number of flips with a fair coin, the expected *ranges* of heads and tails. For ten flips, for example, the statistician might tell us that there is a one-in-ten chance that we might observe as many as 8 heads (or tails) and only 2 tails (or heads). So, if we were to flip a coin 10 times, and find 8 heads and 2 tails, we should conclude not that the coin is unfair, but that there is still a one-in-ten chance it is fair. Nine heads against one tail might, according to the statistician, tell us that there is still a one-in-twenty chance the coin is fair. We could conclude, with either of these outcomes, that the coin is not a fair one, and we have only a small chance of being wrong. Perhaps a better strategy would be to flip more times – create a large "sample." We would learn from the statistician that, with 100 flips, an outcome with 67 heads and 33 tails (or vice versa) has a one-in-ten chance of occurring with a fair coin; now, if we observed 80 heads and 20 tails (identical to the 8–2 split with 10 flips), we are more certain the coin is unfair than we were with 10 flips. Increasing sample size increases certainty.

When examining relative risks the epidemiologist is doing some-thing like testing a coin for fairness, except the outcome concerns the association between two phenomena. A true lack of association between exposure and cancer risk is analogous to a 50–50 split from coin flipping, whereas an association would be like finding an unfair coin (except, of course, there's nothing "unfair" about an association, as long as it's a true one).[1] In the case of the coin, there is a chance – calculable by statistical methods – that we could observe a deviation from 50–50, and still be working with a fair coin. A similar calculation can be performed on the relative risk numbers. Thus, for example, it is possible to estimate the range of relative risks that could, with a specified confidence, still represent 1.0 (analogous to the 50–50 coin split). So, for example, the statistical analysis might show that for a cohort of a specific size, there is a one-in-twenty chance that the relative risk falls between 0.8 and 1.3. If everything else were equal, and the epidemiologist did a study with a smaller-sized cohort, the range for a one-in-twenty chance would be wider, say 0.6 to 1.8; and for a larger sample of subjects, it might be narrower, say, 0.95 to 1.12. As the sample size increases we get closer to the truth about the relative risk. Statisticians call the interval they calculate a *confidence interval*; if it represents one-in-twenty odds, it is called a 95% CI, because it means that if the very same study were to be repeated, with exactly the same population sample size, there is a 95% chance that the relative risk would fall within the specified range (and only a one-in-twenty, or 5% chance, that it would fall outside that range). The statistician can calculate 90% CIs (a narrower range for the same sample size), 99% CIs (wider range), and so on.

The analysis of results from a cohort study is, then, based on a look not at the relative risk alone, but at the CI around it. The choice of CIs to use for evaluation is based on convention, and 95% CIs are most commonly used, but other CIs are often examined.

Let us say the epidemiologist has followed two groups of people exposed to arsenic in drinking water, one group consuming water containing a relatively high level of arsenic contamination and the other consuming water with only a trace level of arsenic (zero is unlikely, since some arsenic occurs naturally). After following the two groups, the epidemiologist finds that the more highly exposed group exhibits a relative risk of bladder cancer of 2.1, with a 95% CI of 1.15 to 3.5. Note that the lower end of the 95% CI is greater than 1.0. This fact

---

[1] In technical terms, the 50–50 split, the absence of an association, is called the *null hypothesis*. An outcome demonstrating an association is one in which the null hypothesis is shown to be incorrect.

allows the statistician to conclude that there is truly an elevated risk of bladder cancer and that there is less than a one-in-twenty chance that such a conclusion is incorrect. That is, the 95% CI of 1.15 to 3.5 means that, with the size of the population sample the epidemiologist was able to assemble, there is far less than a one-in-twenty chance that the relative risk of bladder cancer is as low as 1.0, equivalent to no elevation.

The epidemiologist might want more confidence, and so might ask the statistician to calculate the 99% CI for the result. The 99% CI, according to the statistician, is a range of relative risk of 0.95 to 3.8. Here the lower end is less that 1.0, meaning that there is a one-per cent chance (or slightly more) that the relative risk is not greater than 1.0. So, using a 95% CI, the epidemiologist concludes there is an elevation, but with the more rigorous test, there is no elevation in bladder cancer risk. With the more stringent criterion – which reflects less willingness to be wrong about the outcome – there is no increase in risk (the result is not *statistically significant*); but if the epidemiologist relaxes the criterion, so is willing to accept a somewhat greater chance of being wrong, then an increased risk is found.

These sorts of outcome are not uncommon, and can cause confusion because it is difficult to explain concepts such as "statistical significance," or "not due to chance." The language for expressing probabilities of these types, for elucidating the meaning behind terms such as these, is, to say the least, arcane. The epidemiologist is dealing, necessarily, with degrees of confidence, not absolute confidence. If relative risks are large, say at least a tripling (relative risk of 3 or more), the statistical issues become less important (unless the study involves very small numbers of subjects); but many results fall in the range of one to three, and epidemiologists can have only various degrees of confidence that the relative risks are truly elevated. There are, unfortunately, many examples of cohort studies involving high-profile public health issues that show small and statistically ambiguous elevations in relative risk, and this is one of the reasons why epidemiological outcomes may generate confusion and controversy. The problem is not necessarily in the conduct of the study (though it can be), but rather in the way results are broadcast to the world and interpreted by individuals with different, often opposing, interests and levels of understanding.

Case–control studies do not involve estimates of relative risk – they are initiated with the knowledge that cases exist, and controls are chosen to match the cases in every way possible, with the exception of the presence of the disease being studied. The epidemiologist is looking for exposures or factors, if any, that occur more commonly in the past lives of the cases than in those of the controls. Rather than looking for relative risks, the case–control investigator determines something called an *odds-ratio* (OR) – a figure that is the ratio of the odds that a particular factor or exposure occurred in the lives of the cases, to the odds that the same factor or exposure occurred in the life of the controls. The "odds" are based on actual findings, they are not guesses or statistical figures, and an odds-ratio of 1.0 means that there is no difference between case and control with respect to the occurrence of the exposure in their past lives (or, present lives, if the exposure is a continuing one). Odds-ratios greater than 1.0 suggest an association exists; statistical analyses similar to those used for relative risk are required, and results are subject to the same types of limitations and ambiguities.

Return now to a point made early in this section. Statistical associations, the existence of which cohort and case–control studies can reveal to us, are not, of themselves, evidence of causation. If found, and even if they are *strong* – large relative risk or odds-ratio – they do not establish that the phenomena being investigated, the phenomena that are associated (disease and some exposure or other factor) are causally related. To see most easily why this is so, it is best to examine the difficulties that need to be overcome to move from association to any conclusion about the existence of a causal link.

### Dealing with bias and confounding

That the subjects of cohort and case–control studies are not laboratory animals – genetically homogeneous, all fed exactly the same diets and held in the same environments, all exactly the same age when used in experiments, none hooked on cigarettes or alcohol, none ill or taking medicines – makes for difficulties. Causation could, in theory, be established if groups of individuals could be selected for a study and matched in the same way that lab animals are selected and matched for experiments – if the only difference between two groups so matched is exposure to some substance – and the exposed group develops more cancer of a certain type than the unexposed group (this is the observed association), then it may be highly probable that the association represents a cause–effect relationship. There is, in such an example, only one identifiable difference between the two groups, so the resulting effect – excess cancers – is very likely to have been caused by that difference. (We need to return at a later time to the problem of what causation means in the context of a disease that involves multiple steps, each step possibly influenced by different exposure and factors. For the moment we shall simply note that, in the examples studied by epidemiologists, the "single cause" finding indicates an effect, within the circumstance of the particular population studied, that was of overwhelming influence on disease risk, so powerful that it shows up as "the cause." In fact, even in the population studied, it was likely that there were other factors at work, and that the single factor studied stood out as the most significant and identifiable contribution to overall risk.)

There is always a chance, in epidemiology studies, that the groups studied are not matched on some factor that influences the disease being studied. If that factor is present in one group, and is absent in the other, and the investigator is unaware of that influence, then the observed results are said to be *biased*. As used by epidemiologists, the term does not carry its common meaning, in which it is suggested that the investigator intentionally introduces some factor to ensure a desired outcome. Although deliberate attempts to deceive, to "fudge" results, are not unknown in science, here we are worried about factors that may influence results in ways that are hidden from us, so that we arrive at erroneous conclusions.

Epidemiologists need to work diligently to weed out possible bias, in the selection of subjects for study and in the collection of data on exposure to possible causative factors. If, in a given study, an association is found, but bias is identified and cannot be accounted for, it may be that causal inferences can simply never be drawn.

Confounding is different, and, if it is found, is usually a more significant impediment to reaching conclusions regarding causation. Here the epidemiologist is dealing with the fact that the association found, say between exposure to chemical X and cancer C, while real, is not causal because some other factor (F) is actually the cause. Chemical X *appears* to be the cause, because the epidemiologist has studied X and found an association with C. But it turns out that F (the actual cause of C) is also associated with X (they occur together more frequently than they would if chance alone were acting), but the epidemiologist has studied only the relation between X and C. Drawing the conclusion that X must have caused C would, in the case of this kind of confounding, be a serious error.

Consuming large amounts of coffee is associated with heart disease. But heavy smoking and heavy coffee consumption go together, they are associated. Cohort studies that show an association between high coffee consumption and heart disease can be analyzed to isolate the effects of coffee consumption alone, by focusing only on individuals who do consume large amounts but who do not smoke. The associations found disappear when this is done. This is a clear case of confounding, but one that has been identified, analyzed, and eliminated as a concern.

Smoking is a big confounder in many studies, because it is strongly associated with a number of unusual behaviors and because it is a causative factor in many different diseases. Epidemiologists can usually deal with smoking, but differences in smoking patterns in controls and exposure groups are frequently problematic. Getting accurate smoking histories by asking questions of individuals in case–control studies (or information on any type of behavior that people believe to be "bad for them" or embarrassing, or that they think the investigator might disapprove of) is always problematic.

So, after associations are found, epidemiologists need to double check their study design and findings to reach for sources of bias, and must do everything possible to analyze any such bias to determine whether it could have a significant effect on the observed association. Then, confounders need to be similarly evaluated.

If an association exists, if bias and confounding are judged to be not significant, then the epidemiologist can conclude that the association represents a *cause–effect* relationship. Right? Not quite. There is still more to be done.

### Bradford Hill

In the 1950s and 1960s, Bradford Hill and Richard Doll did great battle with the tobacco industry over the results from their cohort studies of smokers and non-smokers. The associations they identified between smoking rates and excess cancers of the lung were attacked by the industry and even some independent scientists, primarily those who were in the viral etiology camp of cancer causation. Associations maybe, but epidemiology was simply incapable of identifying cause–effect relations. The only reliable way to establish causation was through adequately controlled studies, and no matter how well these studies were done, went the argument, no matter how strong the association, no matter how well confounders were dealt with, uncertainty remained. Epidemiology, on its own, was incapable of establishing cause–effect relationships of any type.

In a way the critics of professors Doll and Hill were relying upon Koch's postulates. Robert Koch (1848–1910) was a German infectious disease specialist who set out, in 1875, a set of criteria, or postulates, that came to be regarded as the best guide to establishing causation in medicine. Koch proposed that several criteria needed to be met. First, he stated that individuals with the disease must be shown to carry the suspected cause – the infectious agent – more commonly than individuals who do not have the disease. At the time Koch was active, the infection theory of disease was ascendant, and it was his own field of study. Second, Koch proposed, it must be possible to isolate the suspect agent; and, finally, it must be possible to reproduce the disease in animals by inoculating them with the suspected agent. These postulates could not be met for cigarette smoking and lung cancer, at least at the time Doll and Hill were reporting their cohort study findings. They may have met the requirements of the first postulate, if you accepted their findings and substituted "smoke chemicals" for "infectious agent," but they had nothing related to Postulate 2, and the only experimental work available showed that extracts from tobacco smoke could cause skin tumors to appear when they were applied to lab animals, in the manner of the early Ishigawa studies and the studies of PAHs extracted from coal tar.

If Koch's postulates could not be fulfilled, the critics argued, the Doll–Hill results could not be accepted as showing causation.

Notwithstanding the critics, the public health community had come to believe, by the early 1960s, that the Doll–Hill findings, together with the case–control results from Graham–Wynder, and several other lines of evidence, were sufficient to support a conclusion that cigarette smoking was a cause of cancer. The Surgeon General of the United States, convened an expert advisory group on the issue, and its report "Smoking and Health" was published in 1964.

Bradford Hill had, as the smoking debate went on, given some considerable thought to the problem of causation in chronic disease. Koch's postulates were perhaps adequate for dealing with infectious disease, where the causative agent remains in the body and is biologically active during the course of the disease, and where the disease occurs almost simultaneously with the onset of infection. Cancer was quite a different creature, especially if it was correct that the disease could be brought about by agents that did damage and left the body, as the damage slowly, very slowly, progressed to disease, perhaps under the influence of other factors. The Koch criteria made no sense in circumstances such as these.

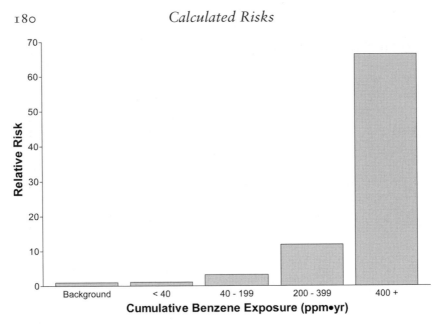

*Figure 6.1  Relative risk of leukemia associated with occupational benzene exposure.*

Bradford Hill proposed that epidemiologists consider a different and, in several ways, more demanding set of factors for judging causality. He noted that, because of the inherent limitations of epidemiology studies, no single study was likely to provide sufficient evidence for causation. Rather, he proposed, causation should be judged according to the weight of evidence derived from several studies. Thus, if similar associations are observed in several studies, especially when they involve different populations, methods of study, and even different investigators, the case for causation would seem to strengthen. Associations that are strong – based on large relative risks or odds ratios – are, for obvious reasons, more compelling than are weak ones. Many times it is possible to study not just one exposed group and compare it with an unexposed group, but to study several groups having different degrees or levels of exposure; if risks are found to increase as exposures intensify, this suggests a causal relationship, because it fits with what biologists understand about dose–response relationships – generally, the risk of adverse health effects increases as exposure level (dose) increases, no matter what chemical exposure is involved (see Figure 6.1).

Another type of biological evidence that supports a case for causation derives from experimental work with animals; if the substance under consideration is carcinogenic in animals, associations seen in epidemiology studies become biologically supportable. Sometimes it is possible to conduct a study after intervening to remove the suspect substance; if risk declines following such an intervention, the case for causation strengthens (recall Pott's chimney sweeps).

The various factors Hill discussed are now commonly used to judge evidence from epidemiology studies. The exercise usually involves groups of experts (and this is best undertaken when experts who have actually conducted one or more of the studies are not involved in the judgment process). In the area of chemical carcinogens the most important of such exercises, as we said at the start of the chapter, are those undertaken periodically under the auspices of the Internal Agency for Research in Cancer. Experts gather to consume Lyon's great cuisine and, between meals, to consume large amounts of epidemiology and experimental data. They judge the quality and limits of individual studies, and then attempt to interpret all the available evidence, using criteria generally similar to those proposed by Bradford Hill back in 1965.

Epidemiology evidence is rated as "sufficient" to establish a causal link to some form of cancer; or as "limited" (a role for bias or confounding cannot be eliminated); or as "inadequate" (insufficient in quality or quantity for evaluation). Some of the substances for which IARC panels have concluded there is sufficient evidence of a causal role in human cancer are listed in Table 5.1.

Another important organization involved in the evaluation of evidence of carcinogenicity is the National Toxicology Program (NTP). In its last report (The Report on Carcinogens Tenth Edition, 2001) the NTP listed 49 chemicals and chemical mixtures as "known to be human carcinogens" (equivalent to IARC's sufficient evidence category). The most recent additions to the NTP lists include beryllium and beryllium compounds, steroidal estrogens, certain compounds of nickel, dioxin, and wood dust. The NTP report lists close to 200 substances as "reasonably anticipated to be human carcinogens," based on some (limited) epidemiology evidence and convincing animal carcinogenicity data.

How this type of information influences the risk assessment and management processes is the subject of much of the last third of the book.

We might in closing note that there are serious epistemological questions, mostly unresolved, that make absolute conclusions about cause–effect relationships in science virtually impossible. The "sufficient" evidence categorizations, and the judgments used to create them, might be seen as imperfect yet useful for practical public health decision-making.

## Laboratory studies

In the years following the Yamagwa and Ichikawa demonstration in 1918 that neoplasms could be produced in experimental animals by long-term application of chemical carcinogens, laboratory scientists all over the globe took up the study of such substances. Some of these scientists, like Kennaway and Hieger in 1920s London, were interested in identifying the specific features of a molecule's structure – the arrangement of and bonding between atoms – that give rise to its capacity to induce malignancies. Others attempted to study the chemical and biological mechanisms that come into play once a chemical carcinogen is introduced into the body or applied to the skin.

Still other scientists worked on the problem of perfecting the use of lab animals as carcinogen detectors. Like Wilhelm Hueper, they wanted to identify carcinogens as efficiently as possible. They did not think it wise to wait until cancers showed up in people, so their causes could be studied by epidemiologists, and were determined to use rats, mice, and guinea pigs, and even dogs and monkeys to learn which chemicals might pose a cancer threat to people. Chemicals identified in lab studies would, if those scientists had their way, be subject to strict regulatory controls.

Of course, many scientists who worked in the laboratory setting were not so sure the results of their work should be taken quite so seriously. They doubted that results from high-dose studies in animals had much relevance to typical human exposures – such experiments would, at best, reveal some of the secrets of the ways carcinogens worked, but were of little use in predicting human risks.

National Cancer Institute scientists took the lead in developing the cancer bioassay protocol. The Institute had, since Heuper's early days, been engaged in an extensive program of testing chemicals for carcinogenicity. The quality of animal tests for carcinogens undertaken in the 1940–1960 period was highly uneven. No widely accepted protocols were available, and so it was difficult to distinguish reliable from questionable results. Criteria for setting the doses to be used in different treatment groups were not much discussed. The number of treatment groups and the number of animals to be used were in no way standardized. Even the criteria to be used by pathologists for diagnosing tumors were poorly characterized; what one pathologist would call a malignancy, another would call hyperplasia (this problem still persists to a small degree).

Of course, we should not think of this early work as somehow "wrong." It was research, most of it. Individual investigators in government and academic labs were exploring the use of animal models. Most were interested not only in the question of whether some substance was carcinogenic; they were also trying to understand how the substance brought about malignant changes. Standardized protocols were of little interest.

Several NCI scientists began, in the early 1960s, to think there was a need for standardized protocols. They thought that one extremely important public health goal was simply to identify, using well-understood animal models, chemicals that had the capacity to induce malignancies. The regulatory and public health community could then decide how and to what extent human exposure to those substances should be controlled.

A number of principles guided the NCI scientists. First, it was clear from much earlier research that repeated, daily exposures for nearly the full lifetime of the lab animals, would be needed to ensure that a substance's carcinogenic properties, if they existed, could be detected. This meant the carcinogen bioassay would extend for about two years in mice and rats, but for much longer times in non-rodent species like dogs and monkeys. For this reason, mice and rats and sometimes guinea pigs and hamsters have been most often used to test for carcinogenic activity.

Other considerations for the design of the cancer bioassay concern the number of animals of each sex to be exposed at each dose level, the number of such dose groups, and the size of each dose. Mode of dose administration is another important consideration. Because it could be expected that only a fraction of the animals in each dose group would develop malignancies (assuming the substance tested turned out to be a carcinogen), and that this fraction would increase as the dose of carcinogen increased, the number of animals selected for each dose level became a strong determinant of the bioassay's capacity to detect a carcinogenic response. Statisticians are concerned, as we have said several times, with sample size – the number of test subjects – and tell us that this has a strong influence on bioassay results. Some aspects of this subject are worth repeating here, because they influence strongly the use of animal study results in risk assessment.

## Sample size and bioassay detection limits

The sample size problem can be understood best with a simple example. Let us suppose we are going to test some chemical, call it X, using lab rats. We intend only a simple experiment, with two dose groups. We have available a supply of young male rats, and from that supply we randomly select 50 animals for the control group and 50 (unlucky) animals for the one test group.

We next pick a dose of X; we need to be sure it is below a level that will harm or kill animals too early – we want these animals to live a very long time to be sure we give X a chance to express any carcinogenic property it might have. We know from early studies how much of X is acutely toxic, and we also know how much is toxic after repeated dosing for, say, 90 days. From this information we make an educated guess about the dose level that will be tolerated by the animals for the long term. We mix X into the rat's chow at a level that, when the rats eat their normal amount of food, will ensure they ingest the targeted dose. If, for example, our targeted dose is 5 mg/day and we know that rats consume 25 grams of feed each day, we need to mix X into the feed at a level of 5 mg for every 25 grams.

We begin. The control animals are held under conditions identical to the treated animals, and get the same feed, minus compound X. We wait. We record body weights daily, and see that the two groups (C for control, T for test group) are gaining and then maintaining similar weights. Weight is important, because as we noted in Chapter 3, one of the first indicators that a toxic process is underway is failure to gain weight in the pre-maturity phase of life, and loss of weight later. Months go by, and all animals appear to be healthy. At 24 months the animals are killed. A pathologist is called in and dissects each animal and looks carefully at every organ and tissue. They note, during what is called the "gross" examination, that everything seems normal in both Groups C and T, but that the livers of 40 of the 50 rats in Group T clearly contain tumorous growths. So, all the livers from both C and T animals are prepared for histological examination; tissues are taken for close examination under a microscope. Another pathologist looks at all 100 (secretly coded) tissue slides, unaware of whether they came from C or T. He diagnoses both benign and malignant tumors, and when the code is broken and the results tallied, they are as follows:

| Group | Total number of animals | Animals with benign liver tumors | Animals with malignant liver tumors | Total animals with tumors |
|-------|------------------------|----------------------------------|-------------------------------------|---------------------------|
| C | 50 | 2 | 0 | 2 |
| T | 50 | 36 | 6 | 40 |

Note that in Group T, the total number of animals with tumors is not the sum of those with benign and malignant tumors; this means only that two of the animals had both benign and malignant tumors.

For the purpose of analyzing these results we decide to ignore the fact that some of the tumors are benign. If we were treating a patient with cancer it would be vitally important to understand the stage of progression of the tumors, but that is not what we are doing here in this experiment. We guess that, with enough time, the benign tumors would progress to malignancies, and so we consider them indicative of a carcinogenic process. National Cancer Institute guidelines developed during the 1970s specified that such a combining of tumor types was appropriate, unless the benign tumors were of a fairly rare type that were known never to progress to malignancy. This practice of combining has been controversial, and until recently was not normally done in many European countries. The argument against combining has pretty much dissipated in the United States.

Our experiment thus shows a tumor incidence of 2/50 (4%) in Group C and 40/50 (80%) in Group T. We call in our statistician and ask whether the incidence in Group T is "statistically significant." "You idiot," he says, "you don't need me to tell you that; it's obvious that these two responses are different. No way could they be statistically indistinguishable."

The statistician has tools available, first developed back in the 1930s by R. A. F. Fisher, a British statistician who undertook to analyze the effect of sample size on experimental results such as these. We have drawn a limited number of animals from what is, in theory, a huge pool of available subjects. If we were to repeat our experiment it is possible, because of chance alone, we could get a slightly different result, say 1/50 in C versus 41/50 in T. And if we were to repeat the experiment again and again and again, we would sometimes see the same result and sometimes see slightly different results. The statistician is able, based on the sample size and some well-accepted statistical assumptions, to predict the role of chance. That role may be expressed as a confidence interval; it is estimated using procedures similar to those used in CI calculations on relative risk measurements in epidemiological studies. In the animal experiment we produce *absolute* not relative risk outcomes, although if we were to divide the risk for group T by that for group C, we would have the relative risk.

A statistician analyzing our results might tell us that the 95% CI on the C response of 2/50 is 0.7/50 to 3.9/50, and the 95% CI on the T response of 40/50 is 37/50 to 44.2/50. This means that if we were to repeat our experiment – the identical experiment – 100 times, in 95 of those experiments the results would fall somewhere within these intervals – there is only a 5% chance the results could fall outside these intervals. The statistician could also calculate the 97.5% CI or the 99% CI, both of which would be somewhat wider than the 95% CI.

If the CI for the C group and that for the T group were to overlap this would mean that we would lose confidence that the responses in the two groups were different – we would have to admit that what appear to be different outcomes are not distinguishable from a purely chance outcome. So, if our experimental result were different, with Group C showing, for example, a liver tumor incidence of 3/50 and Group T showing 7/50, we would definitely need the services of our statistician. They would say that the 95% CI from Group C was 1.1/50 to 5.6/50 and that the 95% from Group T was 4.7/50 to 9.5/50. In this case the lower end of the interval for Group T was below the upper end for Group C. So, at the 95% level of confidence we must admit that these two results, those for Group T and C, are statistically indistinguishable – they cannot be distinguished from a result based purely on chance. Of course, we could elect to "loosen" our criteria for statistical significance. If we were to select a 90% CI, which signals a greater willingness to accept a false positive outcome (one that increases the chances that what we label a carcinogenic response is not correct) we would see that the CI's of C and T no longer overlap. But, using the usual 95% level of confidence as our criterion, we would have to conclude, with this second outcome, the experiment did not provide evidence for the carcinogenicity of compound X.

Notice that we have been forced to this conclusion even though 14% (7/50) animals developed liver tumors! This is a very large risk. The human risk of developing *any* kind of cancer over a lifetime is near 40%, and here we have one chemical producing a 14% incidence in this small group of test animals, and we cannot validly call it a carcinogen. To do so would be to defy fundamental principles of the statistics of chance.

Of course, the greater the number of animals in various test groups, the greater the chances of observing an effect if it is truly present. Another way to say this is that the CI narrows with increasing sample size. One easy way to demonstrate the effect of sample size is seen below.

| Number of animals | Minimum tumor incidence statistically distinguished at the 95% confidence level from a zero tumor incidence in control animals |
|---|---|
| 10 | 4/10 (40%) |
| 50 | 5/50 (10%) |
| 100 | 8/100 (8%) |
| 100 000 | 2/100 000 (0.002%) |

One footnote to the above illustration concerns the incidence in control animals, which is said to be zero animals with tumors. Such an outcome is rare, because even genetically homogeneous, healthy, well-nourished lab animals, who don't smoke, eat excessive amounts of animal fats, or get exposed to sunlight, pollutants, etc., develop tumors over the course of their lifetime. If the control incidence is greater than zero, then the minimum incidence of tumors in the treated groups has to increase to reach significance – recall the earlier example, from our experiment, where 7/50 in Group T was not demonstrably different from 3/50 in Group C; if Group C had been 0/50, then the 7/50 response would have been statistically significant.

Now, finally, the significance of all this. Because, as a practical matter, it is difficult to conduct bioassays with more than 100–150 animals per dose level (usually split evenly between males and females), it can be seen from the above table that in the best of circumstances, with a zero or very low incidence outcome in control animals, it would be necessary for a tested compound to induce something like an 8–15% tumor incidence before we could fairly label it a carcinogen. This is a fairly large risk, yet our typical cancer bioassay has what might be called a limit of detection at about this level. Like chemical assays, bioassays are limited in their ability to detect effects – in this case, the carcinogenicity of a chemical substance.

The various government scientists who, in the 1960s and 1970s, were devising standardized cancer bioassay protocols, recognized the detection problem, and so found a way to compensate for it. The way they found became controversial, and even the subject of much ridicule.

### The maximum tolerated dose

To avoid "false negative" outcomes the government scientists and their academic advisors advocated the use of the maximum tolerated dose of a compound to be tested for carcinogenicity. The maximum tolerated was a dose animals could tolerate without adverse effects that would shorten their lives. Anything greater than this dose would not be tolerated, would shorten lifespan, would reduce the chances of carcinogenicity detection. Anything lower than what came to be labelled the MTD would increase the chance (because of the sample size problem), of missing a carcinogenic effect. In this way, the protocol experts reasoned, the chances of getting a negative result (here "negative" means that the bioassay does not demonstrate carcinogenicity) that was false (misleading) would be minimized. That is, use of the MTD would maximize the chances of detecting a carcinogenic response when the tested chemical was, in fact, carcinogenic. Conversely, the observation of negative results – no excess of tumors – in animals exposed at the MTD would provide a high degree of assurance that the tested chemical was truly not carcinogenic in that test. Such a negative outcome at doses less than the MTD would not be at all reassuring that the tested substance was without carcinogenic activity.

Thus came "high dose testing," animals exposed at hundreds or thousands of times the dose that might ever be experienced by human beings. If positive results were obtained at such high doses, what possible relevance could they have to human health? Indeed, these cancer bioassay protocols, which were adopted by the NCI and other government agencies for their own testing programs, and which regulators such as the EPA and the FDA required industry to pursue (for the cancer testing of substances over which those agencies had authority) became highly controversial. Thus, the goal of the government scientists to avoid "false negative" outcomes was immediately attacked by scientists (and not only industry scientists) who argued that positive results at such high doses were "false positives," meaning they had no relevance to human health.[2] This controversy has been a continuing

---

[2] This is not a proper use of the phrase "false positive." In bioassay terms, a result is said to be a "false negative" when a true positive effect is missed, and a "false positive" when a true negative effect is reported as positive. A cancer effect at an MTD is not a true false

one, although nowadays it takes a more sophisticated form than it did when these protocols came into use.

A couple more points should be made about the MTD. First, it is not easy to identify the MTD. It, of course, varies considerably among chemicals, according to their toxic properties. Chemicals that have high toxicity after short term exposures have MTDs relatively lower than chemicals that have very low short-term toxicity. By "high and low toxicity" we refer to the doses necessary to cause some toxic effects, and a "high toxicity" chemical causes effects at lower dose than a chemical having low toxicity (this makes sense, even if the words are a little confusing). In any event, the selection of the MTD for a particular chemical involves some educated guessing based on the results of short-term toxicity studies, and, especially in the early days, the guesses were sometimes seriously in error (i.e., the MTD was too high and animal survival was poor, or it was too low and possible cancer effects were missed).

Also, the earlier protocols called for not only MTD groups, but also one at half the MTD, so that some idea of the relationship between dose and tumor incidence could be developed. Over the years, protocols have changed to incorporate more dose groups, but the MTD group is still included, although with some improvement in the basis for its selection.

## Dose–response relationships

Dose–response relationships for carcinogens are generally similar to those for other expressions of toxicity. As suggested in the example reviewed above, they are typically plotted as administered dose of carcinogen versus lifetime incidence of tumors at a given site (combined benign and malignant tumors unless, as noted above, there are biological reasons not to do this). As with other toxic responses, the absence of a dose–response relationship for specific tumor types is taken as strong evidence that the chemical does not cause that type of cancer. The lifetime incidence of tumors, expressed as a fraction (number of animals observed to have a specific tumor divided by total number of animals at risk) is called the *lifetime cancer risk* (for the animals).

positive for the rats. Some critics used the term as a short-hand way to argue that it was inappropriate to extrapolate the rat result to people. All assays have measurable rates of false positive and negative results. The ideal assay has minimal rates of both. The government scientists in the case of the cancer bioassay would argue that it was better to tolerate a high false positive rate (calling something a carcinogen when it really wasn't) than a high false negative rate (it could be tragic to miss detecting a true carcinogen).

Table 6.2 *Dose–response relationship for saccharin-induced bladder tumors in rats*

| Dose[a] | Lifetime tumor incidence[b] | Lifetime risk[c] |
|---------|------------------------------|-------------------|
| 0       | 0/50  | 0    |
| 0.1     | 0/50  | 0    |
| 0.3     | 0/50  | 0    |
| 1.0     | 0/50  | 0    |
| 5.0     | 3/50  | 0.06 |
| 7.5     | 12/50 | 0.24 |

[a] Expressed as concentration of saccharin in diets of test animals (per cent of diet).
[b] Number of animals diagnosed with bladder tumors after a lifetime of exposure, with exposures beginning in utero.
[c] Values in center column expressed as a fraction. Only the high-dose response is statistically distinguishable from controls, but it is likely that saccharin also induces the response at the 5% dietary level. The NOAEL is the dose that would be associated with consuming a diet containing 1.0% saccharin, although the actual (unobserved) NOAEL could well be close to 5%.

Table 6.3 *Dose–response relationship for aflatoxin-induced liver tumors in rats*

| Dose[a] | Lifetime tumor incidence[b] | Lifetime risk[c] |
|---------|------------------------------|-------------------|
| 0   | 1/20  | 0.05 |
| 1   | 2/20  | 0.10 |
| 5   | 2/20  | 0.10 |
| 15  | 4/20  | 0.20 |
| 50  | 16/20 | 0.80 |
| 100 | 20/20 | 1.00 |

[a] Dose expressed as parts-per-billion aflatoxin in the diet.
[b] Number of animals observed with liver tumors (all malignant) after a lifetime of exposure, divided by total number at risk.
[c] Values in center column expressed as a fraction. Only the responses at the two highest doses are statistically distinguishable from the control responses, although it is likely that the responses at lower doses are aflatoxin-induced. The NOAEL (on purely statistical grounds) is the dose corresponding to the dietary level of 15 ppb.

Dose–response relationships for two animal carcinogens, strikingly different in potency, are presented in Tables 6.2 and 6.3. The type of information presented in the tables is the usual starting point for risk assessments; as we shall see, human exposures to these carcinogens are very much less than the NOAELs and LOAELs from the animal data.

The two carcinogens differ in carcinogenic potency – the dose required to increase lifetime risk to the same level – by $10^7$ times (ten million). (This calculation requires estimation of the doses incurred by the animals from knowledge of the levels in their diets, which are given in the table, and also knowledge of the amount of diet consumed each day.) Most animal carcinogen potencies fall in a much narrower range between these extremes, with 2,3,7,8-tetrachloro-p-dioxin (Chapter 3) the only compound more potent in animals than aflatoxin. Interestingly, aflatoxin is highly genotoxic, while dioxin appears to have little or no genotoxic activity.

## Too many dollars, too much time

The two-year cancer bioassay involving two species of animals, almost always mice and rats, has become the standard for detecting chemical carcinogens and, with some recent modifications, it remains so. Regulators and public health officials around the world use positive results from such bioassays to assess potential human cancer risk, and to develop regulatory and other public health policies. Results from such bioassays are almost always controversial, although, as we shall see in upcoming chapters, the nature and scientific basis for these controversies have taken some new forms since about the mid 1980s. One aspect of the cancer bioassay that everyone could agree on was its huge cost and time requirements (currently a two-year cancer bioassay involving mice and rats runs close to 1.5 million dollars, and takes about four years from initial planning to final report). Many investigators have tried to develop cheaper and less time-consuming experimental "short-term" tests that might predict long-term effects. To convince the scientific and regulatory communities of the validity and utility of such tests, it was necessary to show that chemicals that were positive in the two-year animal bioassay were also positive in the proposed short-term test, and that a similar correlation held for negative outcomes. This was a tough burden to meet, because many of the short-term tests proposed in the 1960s and 1970s were still costly and time consuming, and most appeared to have some obvious limitations: they might, for example, work well for a particular chemical type or for certain specific cancers, but were not likely to have the more general attributes of the two-year bioassay. None caught on.

Then, in the mid 1970s, Professor Bruce Ames of the University of California at Berkeley came along. We discussed in Chapter 5 Professor Ames' role in the development of tests for genetic toxicity, tests that tell us something about mechanisms of carcinogenicity.

But there is another strand to this story, concerning the use of these tests to identify carcinogens. Ames knew, as did most geneticists, that genes in bacteria are very much like genes in complex organisms, and he further postulated that chemicals capable of inducing mutations in bacterial genes would also be similarly active in humans; assuming those chemicals could enter the human body, and ultimately gain entrance to some cell's nucleus and the genetic material therein. Ames assembled the test, we described in Chapter 5, involving a bacterium with which he had been working, *Salmonella typhimurium*. It was easy to grow the bacterium in culture, and the in vitro test could be run in a single day. He also introduced an ingenious modification to account for the possibility that many chemicals are not carcinogenic (or mutagenic) themselves, but become so when they are chemically modified – undergo metabolism. So, Ames established an assay system that also included an opportunity for the chemical to be exposed in a test tube to the components of liver cells that carry metabolizing (P450) enzymes, and for their metabolites to interact with the bacterial gene.

Ames showed, impressively, in papers published in 1974 and 1975, that a number of substances known to be powerful animal carcinogens in the two-year bioassay were powerful mutagens in his simple test system. Carcinogens, this early work suggested, were mutagens, and the Ames Test is an easy and inexpensive way to detect carcinogens. This work also strengthened the belief that mutation must be a key event, if not the key event, in the carcinogenic process.

During the mid 1970s the Ames test was headline news in the carcinogenesis community and among regulators. But there was a lot of skepticism – could we really trust such a simple test as a way to identify carcinogens and, perhaps more importantly, could we drop concern for chemicals that failed to give a positive response to his test?

The Ames test and other tests somewhat more complex that had been available to the geneticists were soon being applied to at least the "screening" phase of cancer testing. No one was ready to use only those short-term tests, but many labs began to use them to prioritize chemicals for long-term testing – those positive in the test should, it was assumed, be prime candidates for the two-year bioassay. It soon became clear, however, that some animal carcinogens were not mutagenic, and some substances that were mutagenic in these various tests were not carcinogenically active in the "gold standard" animal bioassay. The use of Professor Ames' test, and the others we described in Chapter 5, was soon rejected as a definitive way to test for carcinogens.

But they would stay as screening devices and, more importantly, provide substantial knowledge regarding the various biological mechanisms through which carcinogens act.

## Chemical structure

What has emerged from nearly a century's study of chemical carcinogens is that certain molecular structures appear to signal carcinogenic activity. Many specific members of certain classes of organic compounds, grouped together because of structural similarities, have been shown to be capable of inducing excess neoplasms, thus raising suspicions regarding the entire chemical class. PAHs and aromatic amines have already been seen as suspect chemical types. Representatives of a few more suspect classes are shown here.

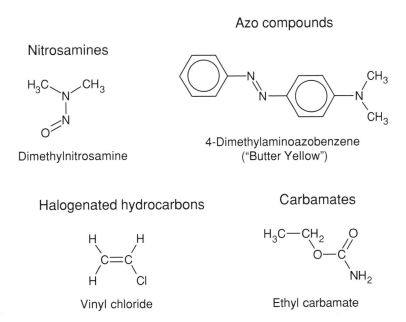

**Nitrosamines**

Dimethylnitrosamine

**Azo compounds**

4-Dimethylaminoazobenzene
("Butter Yellow")

**Halogenated hydrocarbons**

Vinyl chloride

**Carbamates**

Ethyl carbamate

Although any chemical containing an azo group ($-N=N-$) or nitrosamine group ($-N-N=O$) is suspect, not all members of the class turn out to be carcinogenic when tested; this is true for all suspect classes. Sometimes other structural features in the molecule serve to mitigate the effect of the dangerous group, for example, by helping the body to rapidly eliminate the compound and its metabolites.

Many other types of organic compounds have been shown to induce excess cancers. The structures of some of the more interesting of these are depicted here, along with a notation about their origin.

Aflatoxin B$_1$

(mold product, see Prologue)

Safrole

(natural plant product, once used as flavoring agent)

Cyclophosphamide
(anti-cancer drug)

Bis(chloromethyl) ether
(Industrial chemical)

Cycasin
(appears, bound to a sugar molecule, in the cycad nut)

Benzidine
(dye manufacture)

Toxicologists have begun to understand how the chemical structure and some of the other properties of these chemicals contribute to their carcinogenicity. One important part of the effort to reduce reliance upon animal testing involves further validation of the use of chemical structure information along with other properties of a chemical – its chemical reactivity and solubility characteristics in particular – to predict toxicity. The study of quantitative structure–activity relationships (QSAR) is the subject of mainstream toxicology and carcinogenicity research, and has already made its presence felt in some regulatory venues. It might be expected that QSAR, together with information from in vitro studies of several types, will be the future of toxicology.

## Are animal bioassay results to be taken seriously?

Yes, but with caution. Reference has been made several times to the fundamental biological similarities of mammalian species, and to the expected similarities in response to chemical toxicity in animals and human beings. These expectations have been borne out in a large proportion of those cases in which there has been an opportunity to obtain toxicity data in both humans and animals, so that it would be imprudent to ignore the results of cancer bioassays. At the same time these results need to be carefully scrutinized, because they can easily mislead.

All known human carcinogens – the substances ranked by IARC as having been causally linked to human cancers – have been shown to be capable of inducing cancers in some (but not all) species of experimental animals, with the possible exception of arsenic. Arsenic is a human carcinogen, however it has not been adequately tested in animals – so it is perhaps not a real exception to the rule. A few examples of carcinogens that are known to be active in both humans and animals are presented in Table 6.4.

It is of more than a little interest to note that the sites of tumor formation do not always match across species. Benzidine, a substance once widely used in dye manufacture, was shown many years ago to be a carcinogenic risk for the bladder in workers exposed to excessive levels. The rat bladder is not responsive to this substance, but its liver is. It wasn't until Wilhelm Hueper turned to the dog that bladder cancer could be reproduced in a laboratory animal. It is now understood that benzidine metabolism is similar in dogs and people, and that metabolism in the rat takes a different course. It is also understood that certain benzidine metabolites, and not benzidine itself, are the proximate causes of tumors. Knowledge of metabolic differences helps explain the species similarities and differences in tumor response. If we had available the rat data and no human data, we would be in error to conclude that benzidine was a cause of human liver cancer.

Table 6.4 *Some chemicals known to be carcinogenic in humans and their sites of action in animals*

| Chemical | Carcinogenic sites in humans | Carcinogenic sites in animals |
|---|---|---|
| Aflatoxins | liver | Liver in rats, mice, monkeys, and several more species |
| 4-Aminobiphenyl | bladder | Bladder in mice, rats, rabbits, dogs |
| Asbestos (inhaled) | lung, mesothelium | Lung and mesothelium in mice, rats, hamsters |
| Benzidine | bladder | Liver in rats. Bladder in dogs |
| DES | vagina | Cervix, vagina, other tissues in mice, rats, hamsters, monkeys |
| 2-Naphthylamine | bladder | Bladder in hamsters, dogs, monkeys |
| | | Liver in mice, rats |
| Vinyl chloride | liver | Lung, mammary gland in mice |
| | | Liver, kidney in rats |

Empirical information of the type presented above seems to fit theories about the biological similarities of various animal species, including our own, and where differences occur, as with benzidine, it seems that explanations consistent with current understanding of biology are available. So we ask, should we accept as incontrovertible that every animal carcinogen is a potential human carcinogen, when we have inadequate direct information regarding effects in humans? Several hundred animal carcinogens are known to us, and a significant fraction of these can be found in the environment and a larger fraction in the work place. Should all of these be considered cancer threats to humans?

Let us skip by the question of the adequacy of the animal tests used to identify these agents. The general quality of the animal test is obviously of great importance in the overall evaluation and these questions cannot be ignored in the case of cancer bioassays any more than they can in any other type of toxicity test. But the more interesting questions arise when we move beyond the question of study quality.

If we know nothing else than the facts presented thus far – that mammalian species exhibit the same basic biological characteristics (although some differences exist), and that all known human carcinogens are also known to be active in at least one animal species – it would seem foolish to ignore or downplay positive animal tests for other carcinogens, even when no telling human data are available. Indeed, regulatory policy in the United States and the rest of the world embraces animal test data for inferring hazards to humans. But how much of this is science and how much is simply a matter of prudence in the absence of scientific certainty? Surely both are included.

As a simple matter of logic, the fact that all known human carcinogens have been found to be carcinogenic in a least one other animal species is, of course, not proof that every substance found carcinogenic in one or more animal species will be carcinogenic in humans. Logic also informs that even substances found incapable of producing excess tumors in adequate animal tests cannot be absolutely rejected from the class of human carcinogens. We are, for both positive and negative outcomes, dealing with probabilities, not certainty.

What sort of evidence might increase the probability that an agent is or is not a human carcinogen? Although it cannot be proved empirically, it would certainly seem plausible, for example, that a substance producing large excesses of tumors at several sites in several species and strains of test animals and in both sexes, and at multiple doses, is more likely to be carcinogenic in human beings than one that produces only a small excess of tumors at a single site in one species and sex, and that produces no other excesses in other species and strains. Similarly, the greater the number of clearly negative outcomes in animal bioassays, the more convinced we become that the agent is not carcinogenic to humans. This type of weighing of the evidence is one step in the determination of the probability that a chemical is carcinogenic to humans. In character, if not in detail, it is similar to the weighing of epidemiological evidence.

Metabolism data might help. Evidence that a chemical's metabolic patterns in test animals are uniform among several species increases the chances that human metabolic handling of the chemical, if we could obtain information on it, will turn out to match that of animals, whereas the existence of substantial differences among animal species creates uncertainty about which species, if any, humans might match. This business gets complicated quickly, however, because for some chemicals substantial differences in at least rates of metabolism exist among members of the human population.

A particularly controversial aspect of judging the applicability to humans of animal test results concerns the fact that certain sites of tumor formation in particular species, strains, and sexes of test animals are suspected of being uniquely susceptible to carcinogens, or very nearly so, such that excesses at those sites observed in animal tests are considered of dubious relevance to human beings, at least by most experts. Perhaps the clearest example is the kidney of male rats. As the male rat ages his kidney naturally undergoes a predictable series of degenerative changes that seem not to occur in other species, including humans. The female rat kidney also undergoes degenerative changes as it ages, but the changes occur more slowly and are less severe than in males. Certain chemicals such as gasoline (actually a mixture of many hydrocarbons) are capable of accelerating the rate of these degenerative changes in male rats and also of increasing the development of a certain type of tumor in the male rat kidney. These same chemicals produce no such changes in female rats or in either sex of mice. Some biochemists and pathologists believe they understand the underlying biological reasons for these changes leading to tumorigenesis. And they appear to be unique to elderly male rats.

Some other tumor sites are similarly susceptible. The male mouse lungs and liver, for example, tend to develop high and highly variable rates of tumors, even when the animals are untreated with any agent. The reasons for this phenomenon are not entirely clear, although it appears to be due to unusually high populations of certain cells that have undergone initiation into the carcinogenic process. Initiated cells are unusually susceptible to the effects of certain types of chemicals, and progress easily to tumors when assaulted with high doses of those chemicals. The human liver and lungs, as well as these same organs in female mice and in rats, do not seem to contain the same heavy concentrations of susceptible cells, and so may not react so readily to an invasion of chemicals that promote male mouse lung or liver cells to malignancies. We should also mention that the pathologist's diagnoses are not entirely objective – an element of subjectivity enters, especially in the case of rodent liver tumors – so that disagreements about whether a particular lesion is truly a neoplasm arise with surprising frequency.

Many toxicologists are concerned about possible misinterpretation of bioassay results when the MTD (the highest bioassay dose) has turned out to produce serious toxicity as well as a tumor response. They contend that the excessive toxicity that somehow decreased animal survival, or that made them excessively ill, contributed to the production of the extra tumors, and that in the absence of that toxicity, neoplasms would not have developed. In other words the tumorigenic response was not a direct consequence of the chemical, but rather arose from cells so damaged by toxicity that they were put at extra high risk of progressing to the neoplastic state. If human exposure to the chemical were clearly never to reach levels that could cause the overt, initiating toxic damage (and this is almost always the case), then interpretation of test animal results as potentially applicable to humans would be absurd.

These arguments are countered by the point that it is difficult to be sure that what we have called the "initiating toxic damage" was actually responsible for the production of tumors. It might still be the case that the neoplasms would have developed even in the absence of that toxicity. So we should not, so this argument goes, drop our concern until we are certain that cancers would not have occurred without prior toxicity. This is not easy; it is tough to rule one way or the other on this issue without additional data.

As we have stated previously, in these circumstances regulators are usually more fearful of reaching "false negative" than they are of "false positive" conclusions. They prefer to err on the side of safety. The manufacturer whose product is being threatened will obviously object, but is not likely to be successful unless additional data can be brought forth to convince the regulators that exceeding the MTD created a highly artificial circumstance and a false conclusion about carcinogenicity.

A few instances of this phenomenon – high dose toxicity leading to tumors – seem fairly well accepted, even by regulators. Chemical induction of bladder tumors in the rat is sometimes a consequence of the chemical's capacity to produce stones that deposit in bladder tissue. The presence of these solid bodies creates the conditions for the transformation of normal bladder cells to malignancies. If the dose of the stone-producing chemical is dropped below that necessary to create stones, no neoplasms form. The toxic damage – stone deposition – somehow puts the bladder cells at extra risk; the underlying biology of this stone–cell interaction and its relation to carcinogenicity is moderately well understood, and this hypothesis is fairly widely accepted. The action of a non-genotoxic compound that induces bladder tumors in rodents in the presence of stones is probably irrelevant to human carcinogen risk.

But in most cases detailed experimental studies to support the hypothesis for a role of toxicity in production of neoplasia are not available, so regulators rule with caution.

Even if the MTD is not exceeded there can be reasons to worry about positive outcomes obtained at the level. Metabolism is very often a major factor in the production of toxicity and carcinogenicity. A chemical's metabolism may, however, undergo substantial changes, both in terms of the amount of metabolites produced and even in the chemical nature of those metabolites, as the size of the administered dose is changed. In most cases the MTD is estimated from observations of toxicity over a range of doses administered in 90-day studies and ADME studies are not often performed to assist estimation of the MTD. What if, unknown to us, the nature, pattern, or rate of metabolite formation at the selected MTD is radically different from that occurring at much lower doses? Might this not suggest that observation of excess neoplasms at or near the MTD resulted from metabolites that either do not exist or that are formed at much different rates when doses are very low? Even if the high dose metabolic profile results in no unusual toxicity (except excess tumors), such that the survival of the animals is not threatened, might not the high dose results be inapplicable to humans, or even to the same animal species, at low dose where the metabolic profile is greatly different? The incorporation of ADME studies in the determination of the MTD could provide the information needed to circumvent this potential problem, and this practice is becoming increasingly common.

Again, however, regulators become cautious when they are not sure, and they will become convinced that the altered metabolic patterns are significant only with a clear demonstration that the excess tumors would not have occurred in their absence. Such a demonstration cannot be made without additional and usually highly technical studies of metabolism and its relation to dose and to tumorigenesis.

A much longer list of issues relating to interpretation of animal cancer studies could be made and commented upon, but it should by now be abundantly clear that unambiguous results are not common, conflicting scientific interpretations are expected and, in the regulatory setting, most uncertainties are resolved by erring on the side of safety – by a tendency to assume the more pessimistic of two conflicting interpretations. At the same time it should be recognized that there is more agreement among toxicologists on these matters than might be implied from the discussion of the areas of possible conflict. Consensus is no doubt too strong a word to characterize the present state of affairs, but toxicologists do take animal data very seriously; the disagreements usually arise over the highly important details of study interpretation, and not over basic principles.

As a general matter, animal carcinogenicity data obtained from well-designed and conducted bioassays, are used routinely by regulators to assess human risk. In the absence of epidemiological information it is generally not possible to claim knowledge of the type of human cancer that is being assessed (for reasons discussed above), but that is not of particular concern in the regulatory context.

We now turn to risk assessment, and the practical application of the science discussed in these first six chapters. We start off with a chapter that focuses on certain principles that have come to be widely accepted and then move to more practical applications. Some of the newer trends in risk assessment are highlighted in Chapter 9. A final chapter on risk assessment pertains to the interesting and increasingly important problem of its application outside the regulatory and public health arenas, in particular in relation to the legal settings in which specific individuals bring claims of personal injury (toxic harm) resulting from chemical exposure against manufacturers, or other parties said to have been responsible for their exposure.

The quartet of risk assessment chapters is followed by a look at the risk management practices of regulators and the ways in which risk assessment results are used. As is expected, the concluding chapter looks to the future.

# 7

# Risk assessment I: some concepts and principles

Consider the following situations and the questions that arise as a result.

(1) Individuals working in a petroleum refinery are routinely exposed, over the course of an eight-hour work day, to the volatile hydrocarbon benzene, a constituent of petroleum that has been established through epidemiology studies, in quite different occupational situations, as a cause of human leukemia. Is it possible to understand whether and to what extent these specific, unstudied refinery workers are at risk of developing leukemia?

(2) Because benzene is a constitute of gasoline and some other fuels, and because it is so volatile, virtually all of us are exposed, almost continuously, to a certain level in the air we breathe. In some cases, because of fuel leaks and spills, ground waters and surface waters have become contaminated with benzene. What can be said about the threats to health, if any, we all face from these relatively low levels of exposure through our environment?

(3) A compound called methylmercury comes to be present in fish because of various industrial releases of mercury to the environment, and even through its migration from some natural ores present in the seabed. The released mercury undergoes a chemical conversion brought about by the natural biochemical processes of microbes, and becomes methylmercury. The latter compound is readily taken up into fish and because, unlike the inorganic form of mercury from which it derives, it is fat-soluble, it tends to accumulate to a greater degree in oilier species. In some (but not all) studies in certain human populations with relatively high levels of methylmercury intake, children born of fish-consuming women exhibit a variety of more or less serious neurological and learning deficits. Are

such deficits to be expected in children of the vast majority of women of child-bearing age who are exposed to lower levels of methylmercury through this source?

(4) A manufacturer seeks approval from the FDA for the marketing of a new non-caloric sweetening agent. The manufacturer has conducted extensive animal toxicity testing on this new food additive, and has also provided to the FDA information about the chemical's use rates in foods and the expected rate of intake consumers might experience. Is it possible to predict whether the new additive will pose a health risk to consumers if it were to be approved for use in food?

(5) An epidemiology study involving 250 patients has found a significantly increased risk of a kidney disorder in a population using a certain medication that is known to reduce the risk of heart disease. What can be said about the risk of kidney disease in the entire population of nearly 500 000 users of the drug?

All of these questions are about risks to human health resulting from past, current, and (in the case of the food additive) future exposures to chemical substances. They seem to be highly important questions, and if we are ourselves members of one or more of the exposed populations, we would press scientists and physicians in our public health and regulatory institutions for answers to them. And, if those in authority in those institutions are doing their job, they have programs in place to provide the answers.

We would, of course, want more than answers to these questions if those answers revealed that our health or that of our children was indeed in jeopardy: we would want public health and regulatory officials, using whatever legal authority they have, to take action to reduce or even eliminate these risks, and some of us with an activist bent would find ways to exert pressure on the industrial concerns that are the source of the risk. And we would even expect that responsible manufacturers would want to take action to mitigate risks without having to be pressured. In the case of the new food additive, if it were found to be likely to pose excessive health risks, we would expect the FDA to prevent its marketing. The situation involving the heart medication would seem more complex than the others, in that we would expect the responsible regulatory agency, again the FDA, to balance the risk of kidney disease associated with the drug's use against the risk of heart disease that might arise in patients should the drug be banned. Many questions may have to be considered in this type of difficult balancing act (for example, are there other safer and equally effective medicines available?), and the FDA often resolves its part of the problem by ensuring that prescribing physicians are fully informed about drug risks and benefits, and about how that information can guide decisions about individual patients. (Recent controversies surrounding the use of certain pain-killers and risks of heart attack and stroke reveal that either these balancing acts are extremely difficult, or that the FDA and regulated companies are not sufficiently vigilant, or both.)

The types of activity described in the last paragraph are different from those associated with answering questions about risk. The set of evaluations necessary to identify and measure risks to health is carried out within a systematic framework that is called *risk assessment*. The actions needed to reduce or eliminate any risks that are found, through the assessment process, to be unacceptably large, fall within the domain of *risk management*. One can envision risk assessment as an activity that occupies a position between *scientific research* on the adverse health effects of chemicals and on the extent of human exposure to them, and the process of making decisions regarding how to manage risks to health so that people are protected. *Risk assessment* is needed because it is almost never the case that individual sets of research results, whether from the efforts of epidemiologists, the laboratories of toxicologists, or the activities of the scientists and engineers who study human exposure to chemicals, provide information that is useful and of direct relevance to risk management. Risk assessment involves the integration of diverse, and sometimes contradictory, scientific information, often of highly varying quality, to provide descriptions of risk that are (we hope) consistent with the total body of underlying scientific evidence, and that are also useful for practical public health, regulatory, and corporate decision-making.

The remaining sections of this chapter are concerned with the scientific difficulties encountered in the practice of risk assessment – in fact, it will be seen that there are critical aspects of the risk assessment process that cannot be adequately dealt with because of limitations in scientific understanding. Following this chapter is another on risk assessment, devoted to its practical applications, and then comes a third chapter providing examples of some new risk assessment challenges and approaches. After a final brief chapter on risk assessment in the courtroom, risk management returns in Chapter 11.

## The 1983 "red book"

The linkages between research, risk assessment, and risk management, and the scientific and policy features of each, were first systematically set forth in a study by a Committee of the National Research Council – National Academy of Sciences, released in 1983, in a relatively slim volume with red covers. The National Academy is still selling this "red book," and demand for it remains high. The committee that released the report (and of which I was happy and honored to be a member) was assembled by the academy at the request of the US Congress. It was asked to examine various scientific practices and developments within the federal regulatory agencies which were charged under a variety of statutes with regulating potentially toxic chemicals – the EPA, the FDA, the OSHA, the CPSC, and the FSIS/USDA. During the mid-to-late 1970s some regulatory practices, particularly those associated with the assessment of risks from carcinogenic chemicals, had engendered enormous controversy, both within the regulated communities, who thought they were being unfairly skewered by over zealous risk assessors in the regulatory agencies, and the various communities of environmental and consumer activists who believed regulators were insufficiently aggressive in enforcing their legal mandates. Much of the debate concerned basic questions of public policy, and we shall leave discussion of this matter to the chapter on risk management.

But the effort of the National Academy committee was devoted primarily to scientific issues: the alleged distortion of science by government risk assessors, to ensure that risk managers received the answers they wanted, so that they could decide to regulate, or not to regulate, depending upon how they perceived the social, economic, and political pressures under which they were operating. Whether such distortion of science truly occurred was not addressed by the committee, but its many recommendations were designed to remedy certain regulatory practices that at least increased the likelihood of distortion for political purposes. The committee also recognized that a lack of clarity regarding basic concepts and even in definitions of terms was the cause of much of the confusion and contention that plagued the regulatory process.

Figure 7.1 is adapted from the "red book" (actual title: "*Risk Assessment in the Federal Government: Managing the Process*"). The committee offered this figure as a depiction of the broad "framework" under which the three major activities necessary to protect public health from the hazardous properties of environmental chemicals (very broadly defined) – research, risk assessment, and risk management – should be organized. The committee further emphasized that the three involve quite discrete sets of analytical undertakings, and serve different purposes, so that efforts should be made to reduce the chance of inappropriate influence of one upon another. Thus, for example, risk

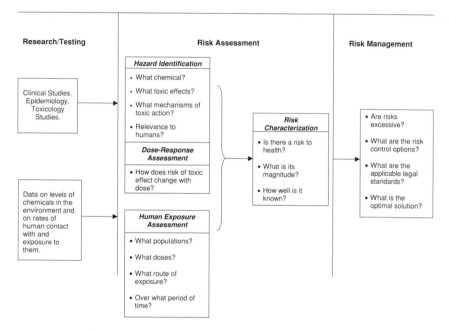

*Figure 7.1   Risk assessment and its relationship to other risk analysis activities.*

assessment should be carried out under its own standards of evaluation, and risk managers should not attempt to alter those standards in specific cases simply to ensure that some pre-determined management objective is more easily achievable. At the same time, the three activities are necessarily linked, and close communications among those involved are critical to efficient public health protection. Thus, within the recommended framework, conceptual distinctions are critical, while isolation of those involved in the three activities from each other would be a recipe for failure.[1]

Within the framework depicted in Figure 7.1, the content of risk assessment proposed by the committee is shown as comprising four analytic steps: *hazard identification, dose–response assessment, human exposure assessment*, and a final, integrating step called *risk characterization*. These four terms and the activities they describe have come to be widely accepted within the risk assessment community, on

[1] One question put to the "red book" committee by the US Congress had to do with the advisability of separating risk assessment practitioners, centralizing them in a single government agency, so that they would be shielded from the regulators and policy-makers. The committee rejected the proposal.

a worldwide basis. While some institutions have seen fit to alter some of these terms, those differing terms nevertheless describe the same basic activities.

To illuminate the four steps of risk assessment in the simplest possible way, we might describe their respective contents and the logical connections among them as follows.

Scientists skilled in epidemiology, toxicology, and related disciplines collect and evaluate all of the scientific literature containing information regarding the types of toxic effect the chemical under review has been shown to produce. Toxic effects include one or more of the many manifestations of toxicity described earlier in this book. The list of adverse health effects produced by the chemical are said to constitute its toxic hazards, and the critical review and evaluation leading to the list is the *hazard identification* step. A discussion of the extent to which causal associations with human disease or toxic harm have been established is an important aspect of this step.

As has been emphasized so many times in the preceding chapters, these various manifestations of toxicity all display dose–response characteristics, where by "response" we refer to the incidence or severity of specific adverse health effects. As we demonstrated in earlier chapters, toxic responses increase in incidence, in severity, and sometimes in both, as dose increases. Moreover, just below the range of doses over which adverse effects can be observed, there is usually evidence for a *threshold dose*, what we have called the no-observed adverse effect level (NOAEL). The threshold dose must be exceeded before adverse effects become observable (Chapter 3). Deriving from the literature on toxic hazards, descriptions of the dose–response relationships for those hazards comprise the dose–response assessment step of the four-step process.

While the hazard identification and dose–response assessments are underway, another group of scientists is at work identifying the population(s) exposed to the chemical, the routes by which it enters the body, and the amount of chemical entering per unit of time (which might be a few minutes, an eight-hour work day, or a full 24-hour period). This *exposure assessment* is thus geared to describing the *dose* incurred by the exposed population (Chapter 2). Knowledge of the duration of time over which exposure occurs (hours, days, weeks, months, years) is an additional component of the exposure assessment step. With the completion of this step, we have sufficient knowledge to complete the risk assessment. Thus, we simply examine the dose of the chemical incurred by the exposed population, and compare it with the dose–response information we have compiled on the toxic hazards of the chemical. This comparison may tell us that the human exposure is below the threshold dose, so that we can conclude the population is not at risk of toxicity. If the population's dose is located somewhere on the dose–response curve above the threshold, then we can conclude that the population is at risk of toxicity, and the size of that risk will depend upon how far up the dose–response curve the population dose sits. This last step, in which the dose incurred by the human population is set against the hazard and dose–response data available for the chemical, was termed *risk characterization* by the academy committee, because it was thought that it should consist of a relatively detailed description of the nature of the hazard to be incurred, its seriousness, and the likelihood of its occurrence (which is gauged by the location of the incurred population dose on the dose–response curve).

As risk assessment has just been described, it can be seen that it draws upon all of the types of scientific studies and research that have been described in the previous chapters of the book. Information and knowledge from those studies comprise the scientific content of the three steps of risk assessment that must be completed before the risk characterization step can be pursued.

The remaining sections of this chapter are devoted to revealing why the outline of risk assessment just presented, while offering a generally correct description of the content of the four steps, is guilty of serious sins of omission. Some of the types of problem encountered in any attempt to complete a risk assessment have been hinted at in earlier sections of this book. It is time to give a comprehensive summary of the problems that need to be overcome during the risk assessment process. Keep in mind that risk assessment does not create new knowledge or information. It is rather a framework within which existing knowledge and data are organized and evaluated for purposes of decision-making. Risk assessment cannot be expected to compensate for lack of knowledge. It might be described as not only concerned with describing risks to human health, but also with making clear what is unknown or uncertain about those descriptions. Research is the only road to acquiring the information or knowledge necessary to reduce these uncertainties.

The strength and quality of the scientific evidence regarding chemical toxicity, dose–response characteristics, and human exposures varies enormously among chemical products and pollutants. Some of the reasons for these differences were described in the chapters concerned with identifying toxicity. The problems encountered in the conduct of a particular risk assessment will be heavily dependent upon the quantity and quality of information available on the chemical under review. So no single account of the problems encountered in the conduct of a risk assessment will be adequate to describe the full universe of possible problems. For our purposes, we shall assume the existence of a moderately comprehensive set of investigations, and attempt to give an account of the difficulties most commonly encountered in the conduct of a risk assessment.

## Scientific obstacles to risk assessment

Scientific evidence concerning toxic hazards and their dose–response characteristics for a particular substance is collected under one set of conditions (call it condition **A**), and is to be used to assess risks that might arise under different conditions of exposure to that substance (call this condition **B**). In some cases, the differences between conditions **A** and **B** are relatively small, but in many cases they are large. Risk assessment necessarily entails *extrapolation* from observations made under condition **A** to allow inferences to be made regarding what might be expected under condition **B**. Here are the major reasons why extrapolation is necessary.

(1) If epidemiological investigations are the source of the data used to identify a toxic hazard and its dose–response characteristics, they almost always involve only a small subset of the total number of individuals exposed. Thus, if we refer to the example of the heart medication set out in the opening pages of the chapter, we see that only a small subset of the total number of individuals using the medication was studied. Given the observation of a certain excess rate of kidney disease in that subset, how confident can we be that the same rate is to be expected in the total population of medicated individuals? What additional information is necessary to ensure confident extrapolation from human population A (the population studied) to population B (the total population of users)? Once we have ventured to engage in the necessary extrapolation, how should the limitations in our understanding of kidney disease risks in population B be described?

(2) In the example of the heart medication, all individuals are probably given identical doses of the drug. But consider the women of child-bearing age who consume fish but incur lower levels of methylmercury intake than the women whose children were studied epidemiologically, and who were found to suffer neurological or learning deficits. Thus, in this case, extrapolation is necessary not only from a subset of the entire population of

fish-eating women of child-bearing age, but also from a highly exposed to a less highly exposed population. We either find a way to make these extrapolations, or we remain silent about methylmercury risk in the general population. (We might simply conclude that there is likely to be "some risk," but such a vague, qualitative statement gives public health officials and regulators little real guidance.)

(3) In many cases, epidemiological data used to identify toxic hazards and dose–response characteristics are collected in occupational cohorts (condition **A**, which could represent the studies of benzene and leukemia, mentioned in the first example). If condition **B** involves other occupational cohorts, as in example 1, then the problems of extrapolation are likely to be similar to those two just described. But suppose condition **B** is the general population, as in example 2? The general population consists of individuals having characteristics substantially different from the occupational cohort that was studied. Condition **A** consists of a relatively healthy working age population; in some industries most of the workers will be male. Condition **B** consists of exposed infants and children, pregnant women, the aged of both sexes, and people who suffer from many medical ailments. Moreover, exposure to benzene may occur not only through the air (as in condition **A**), but also through ingestion of water. And, perhaps most important, the level of benzene exposure associated with condition **B** is likely to be hundreds or even thousands of times lower than that at which excess leukemias are observed (condition **A**). Here we are required to extrapolate from a relatively uniform population to a highly diverse one; from one route of exposure to a second; and from relatively high dose to very low dose. The risk assessor needs lots of basic biological understanding to navigate these extrapolation pathways with anything approaching scientific rigor.

(4) There are many circumstances in which the only information we can develop on toxic hazards and dose–response relationships derives from experiments on laboratory animals. The example of the food additive, presented in the opening pages, is just one of many circumstances in which condition **A** involves animal toxicology data, and condition **B** involves a human population, almost always exposed at small fractions of the dose used in animals, and sometimes exposed for much larger fractions of their lifetime than the animals, and even by different routes. Extrapolations under these circumstances should cause individuals trained in the rigors of the scientific method to seek some form of psychological counsel, or, better yet, to return to the laboratory.

As if these four sets of circumstances did not signal enough trouble, there are other barriers to successful risk assessment. Most have to do with what we have been referring to as the "conditions" under which toxic hazard and dose–response information have been collected (**A**), and the conditions (**B**) under which the population that is the subject of the risk assessment is exposed to the substance of concern. Because the conditions studied, if the investigators have been careful and thorough, have been well characterized, the important factors affecting any health risks observed should be relatively well-established. A well done laboratory experiment, with the necessary controls, will, if it represents condition **A**, be the most readily characterizable set of circumstances. An epidemiological investigation, as it has been described in Chapter 6, has more significant limitations regarding the characterization of the exposure conditions under which individuals have been studied. The problem epidemiologists often have in identifying the exposures and resulting doses individuals in these studies may have incurred probably heads the list of uncertainties. So, in many cases, having firm knowledge regarding condition **A** is problematic.

The difficulties associated with describing condition **B** are usually even greater, because this condition involves large populations whose exposures and resulting doses often cannot be known with high accuracy. These doses are estimated, as outlined in Chapter 2, by combining knowledge regarding the environmental media through which exposure occurs, the estimated or measured concentrations of the substance of concern in those media, the rate of intake of those media into the body, and so on. All of these exposure assessments involve extrapolation from data obtained from limited sampling. Extrapolation and other forms of estimation needed to describe both conditions **A** and **B** further bedevil the risk assessment process. Of course in situations involving pharmaceuticals or food additives, for example, where exposures are carefully controlled, describing condition **B** is not nearly so problematic.

The situation becomes bleaker still when we recognize that, for a given substance, there may be many different "conditions **A**" under which its adverse effects and their dose–response characteristics have been investigated! Results may be available from several different epidemiological studies, in different groups exposed under different circumstances, and with results that are not entirely consistent with each other. Some of the conditions may involve experimental data, similarly variable in outcome and in how they will be interpreted by different scientists. So, when we are faced with toxic hazard and dose–response data from studies involving conditions **A1** through **A12**, which, if any, are most useful and relevant for extrapolation to condition **B**?

At this stage it may appear that completion of the apparently simple four-step process of risk assessment, as depicted in Figure 7.1 and described earlier in the text, is not at all simple, and requires a significant number of extrapolations, or inferences, beyond the data that are typically available. It is often the case, in many areas of scientific endeavor, that inferences are made beyond what has been directly measured. Statisticians can estimate, using well-established tools, the confidence we can have in drawing inferences for large populations from data collected in certain subsets of individuals, and some important aspects of the problems described in the foregoing can be handled in this way. If, for example, chemists have sampled and analyzed certain animal tissues for the presence of dioxins, using a carefully developed and statistically sound plan, then statistical analysis can be applied to the results, and inferences regarding larger (unsampled) sets of animal tissues can be drawn, with some indication of the confidence that can be assigned to the inferred (extrapolated) results.

The problems outlined in this chapter are, however, not all amenable to resolution with this type of statistical analysis. They are not so much problems of statistics, rather they are problems regarding basic biological knowledge and understanding.

Typically extrapolations of many kinds are necessary to complete a risk assessment. The number and type of extrapolations will depend, as we have said, on the differences between condition **A** and condition **B**, and on how well these differences are understood. Once we have characterized these differences as well as we can, it becomes necessary to identify, if at all possible, a firm scientific basis for conducting each of the required extrapolations. Some, as just mentioned, might be susceptible to relatively simple statistical analysis, but in most cases we will find that statistical methods are inadequate. Often, we may find that all we can do is to apply an assumption of some sort, and then hope that most rational souls find the assumption likely to be close to the truth. Scientists like to be able to claim that the extrapolation can be described by some type of model. A model is usually a mathematical or verbal description of a natural process, which is developed through research, tested for accuracy with new and more refined research, adjusted as necessary to ensure agreement with the new research results, and then used to predict the behavior of future instances of the natural process. Models are refined as new knowledge is acquired.

Some will be discarded altogether when they are found to provide inadequate descriptions of the natural phenomena under study. Extrapolations necessary to complete risk assessment may thus involve the drawing of statistical inferences or the use of assumptions and models that have at least tentatively satisfied certain criteria of validity. We need to step through the specific types of extrapolation needed to complete a risk assessment to give concreteness to these somewhat vague ideas. When this is done, in the next chapter, it will be seen that for many of the critical extrapolations needed, there are several assumptions and models that might be applicable, that there is no firmly established scientific basis for discriminating among them, and, when applied, they lead to quite different predictions of risk. The National Academy's "red book" committee recognized this problem, and also noted that its existence provided the opportunity for case-by-case manipulations of risk assessment results to achieve predetermined risk management objectives ("I am worried about having to impose strict regulations on this important pollutant, so select your extrapolation models to ensure the outcome of the risk assessment is not alarming in any way," says the fearful risk manager to the nervous, but compliant risk assessor, who is trying to hold on to her job).

## "Defaults"

The red book's recommendations regarding the conduct of risk assessment contained the following remedies for this potential and serious problem, and for ensuring in other ways that risk assessments are conducted with maximum adherence to whatever scientific knowledge is available, and with a consistent approach to dealing with the uncertainties in that knowledge (many of which relate to the need for extrapolation).

Specifically, the committee urged the regulatory agencies to develop sets of guidelines for the conduct of risk assessments. These guidelines would describe the scientific basis for each of the steps of risk assessment set forth in Figure 7.1, and would also describe how the necessary evaluations would be carried out for each step. The guidelines would also describe the scientific basis and methodologies (assumptions, models) that might be appropriate for the conduct of any needed extrapolations. It would be recognized that scientific understanding is limited with respect to many types of extrapolation, and that several different methods may be equally justifiable for each required extrapolation. To ensure that risk assessments can be completed, and to ensure that the problem of case-by-case manipulation of results is avoided, agencies were asked to select, from the available extrapolation methods, one that would be consistently applied in all cases. The committee recognized that such a selection, given limited scientific understanding and justification, would not be strictly scientific in nature, but would necessarily involve an element of policy. The committee termed these selections "science policy," to distinguish them from the type of policy choices needed for risk management. The regulators were asked not to deviate from these science and science policy guidelines unless it were to become clear, in some specific case, where relevant research had been conducted, that the usual science policy choice was incorrect. In such a case, results from the research could be used instead of the usual policy choice.[2] Consistent use of the guidelines would not guarantee the accuracy of risk assessment results (there is no way to ensure this, because most risk assessment results are not testable in the usual scientific sense), but it would come close to ensuring that risk assessment results obtained for different chemicals can be compared, because they are all derived using the same sets of models and assumptions.

The US EPA has been the most diligent of the regulatory agencies in following the red book recommendations, and has developed, and continues to refine, guidelines for the conduct of risk assessments. Other US agencies have put much less effort into the guideline development process, but the risk assessment procedures they follow generally match those of the EPA. Extensive guidelines have been and are being developed in the European Union. References for these various guidelines are to be found in the "*Sources and recommended reading*" section of this book.

The "science policy" components of risk assessment have led to what have come to be called "default assumptions." A "default" is a specific, automatically applied choice, from among several that are available (in this case it might be, for example, a model for extrapolating animal dose–response data to humans), when such a choice is needed to complete some undertaking (e.g., a risk assessment). We turn in the next chapter to the conduct of risk assessment and the ways in which default assumptions are used under current regulatory guidelines. We might say we have arrived at the central subject of this book.

---

[2] Research might be able to answer questions about the forms of extrapolation having scientific support in the case of specific chemicals, but the development of science-based forms of extrapolation that can be applied generally is not yet possible – hence the continued need for science policy choices.

# 8

# Risk assessment II: applications

I believe it was Mark Twain who quipped, when asked what he thought of the music of Richard Wagner, that "It's not as bad as it sounds." Risk assessment might be similarly described.

Some risk assessors, in apparent disregard for the uncertainties associated with interspecies extrapolations and identifying dose–response relationships, have in recent years been making one announcement after another on the health risks associated with chemicals in the environment. "There is a one-in-one hundred thousand chance that ALAR will cause cancer in children consuming apple juice." "Methylmercury contamination of seafoods leads to 80 000 cases each year of learning disabilities in children." "Dioxin contamination of animal-based foods causes cancer in one of every 500 000 people." The media offer such revelations with increasing frequency, usually accompanied by statements from regulatory agencies designed to quell public fears, remarks from manufacturers to the effect that risks have been greatly exaggerated, and professions of outrage from critics of both the regulatory and industrial communities.

It should be obvious by now that statements about risk of the type cited above, if standing alone, are misleading; if it is not obvious, it will be by the end of this chapter. There are no means available to identify these types of risk with the degree of certainty suggested by the language used. Perhaps the best a risk assessor might do, given today's knowledge, is a summary that goes something like this:

Difluoromuckone (DFM) has been found to increase the risk of cancer in several studies involving experimental animals. Investigations involving groups of individuals exposed in the past to relatively high levels of DFM in their workplace have not revealed that the chemical increases cancer risk in humans. Because these human studies could not detect a small increase in risk, and because there is a scientific basis for assuming results from animal experiments may apply to humans, exposure to low levels of DFM may create an increase in risk of cancer for people. Because of the limitations of animal studies the specific type of human cancer cannot be predicted. The magnitude of this risk is unknown, but probably does not exceed one in 50 000. This figure is the lifetime chance of developing cancer from a daily exposure to the highest levels of DFM detected in the environment. Average levels, which are more likely to be experienced over the course of a lifetime, suggest an upper limit on lifetime risk more like one in 200 000. These risk figures were derived using scientific assumptions that are not recognized as plausible by all scientists, but which are consistently used by regulatory scientists when attempting to portray the risks of environmental chemicals. It is likely that actual risks are smaller than the ones cited above; larger risks are not likely but cannot be ruled out. Regulators typically seek to reduce lifetime risks that exceed one in 100 000. Note that the lifetime cancer risk we face from all sources of cancer is about 1 in 3, so that, even if correct, the DFM risk is a minor contributor to the overall cancer problem. Prudence may dictate the need for some degree of risk reduction for DFM in the environment.

This statement could no doubt be much improved upon, but, based on what we have said in the last chapter, it is certainly much closer to what risk assessors know than those cited earlier.

## An emerging discipline

Like those of toxicology, the foundations of the risk assessment discipline were laid in several different areas of study, and these have begun to merge only within the past two decades. One of these foundations can be located in the work of radiation biologists and health physicists who began, not long after the discovery of radioactivity at about the turn of the century, to investigate the adverse health consequences of exposure to this form of energy. Their work received a major impetus from the development, production, and deployment of nuclear weapons and the use of nuclear materials for energy production; all these activities created opportunities for human exposure to various forms of radiation, the most intensive and widespread of which were those incurred by survivors of Hiroshima and Nagasaki. Some of the models proposed for assessing low dose risks from chemical carcinogens have been in use by radiation scientists for several decades, although the scientific merits of these various models are still the subject of much debate.

Another important foundation for risk assessment can be found in the work of safety engineers. For several decades, concerns about the safety of large physical structures and complex manufacturing and energy-production facilities – dams, nuclear power plants, chemical-manufacturing facilities, and so on – have prompted analyses of the risks that they may fail to operate as planned. Failure analysis, as it is sometimes called, involves assigning probabilities to various events that may lead to a failure – the release of a highly toxic chemical to the atmosphere, for example – so that construction and operating procedures, and various "fail-safe" mechanisms, can be built appropriately into the system. Safety engineers are, of course, also involved in the production of hundreds of types of complex manufactured goods, the failure of which could lead to injury.

The Society for Risk Analysis was organized in 1980 by a group of scientists and engineers from these various disciplines, and toxicologists were included among them. These individuals believed that, whether the issue was the failure of a nuclear power plant or brakes on an automobile, or human exposure to chemicals or to radiation, they were united by a common interest in the analysis of risk. The Society and its journal *Risk Analysis* have prospered and continue to draw new membership. A meeting of the Society brings together an odd but interesting assemblage of engineers, health scientists, statisticians, toxicologists, physicists, molecular biologists, radiation experts, regulators, and even social scientists and psychologists interested in problems of risk perception and communication. More recently the Society has attracted scientists concerned about pathogenic microorganisms and the threat they pose to food and water supplies, infectious disease experts, and those interested in the risks of under- and overnutrition. Terrorist threats have greatly increased interest in risk assessment. It appears that risk analysis is here to stay, and risk analysts stand ready to explore the threat of just about any aspect of modern technology and the natural world.

Risk is the probability that some harmful event will occur. What is the probability that certain types of cancer will develop in populations exposed to aflatoxin in peanut products or benzene from gasoline? What is the likelihood that workers exposed to lead will develop nervous system disorders?

Because it is a probability, risk is expressed as a fraction, without units. It takes values from 0 (absolute certainty that there is no risk, which can, of course, never be shown) to 1.0, where there is absolute certainty that a risk will occur. We say that the lifetime cancer risk from carcinogen A at an average daily dose of B is one in 100 000 (0.000 01); if this number is accurate, it means that one of every 100 000 people exposed to carcinogen A at a lifetime average daily dose of B will develop cancer over a lifetime. The probability also describes the extra risk incurred by each individual in that exposed population.[1]

People are more familiar with expressions of risk associated with various activities than they are with risks associated with chemical exposure. We speak, for example, of the annual risks of dying as a result of certain activities. In the United States, the annual chance of dying in automobile accidents for people who drive the average number of miles is about one in 4000. The average cyclist faces an annual risk of death from pedalling of about one in 30 000. Pack-a-day smokers who began at age 15 incur a risk of death from lung cancer of one in 8. The lifetime risk of developing cancer in the United States is qreater than one in 3, if we include smokers.

These types of expressions of risk are more familiar to people, but they mean roughly the same thing as those described earlier for the risks of toxicity from chemical exposure – with at least one exceedingly important difference.

Information on death rates from automobile or other types of accidents or activities is generally much more solid than that pertaining to most chemical risks. Statistical data, compiled by actuaries, are used to derive such risk information. There is uncertainty associated with these actuarial figures, but most are fairly reliable. Most of the risk information about various cancers, presented in Chapter 5, is of this type.

Most of the risks estimated to be associated with environmental chemical exposures are much less firmly established. So, although chemical risk information is often expressed in the same form as that based on directly measured risks, it is derived using quite different methods, and almost always includes extrapolations beyond measured risk data.

Another important source of confusion in the use of the term risk needs to be re-emphasized here. When a risk assessor states that exposure to DES increases the risk of certain cancers in women, they mean

---

[1] In Chapter 6, in the section on epidemiology, risks were presented in *relative* terms, the ratio of the risk observed in one population to that observed in another. Such relative risks are important and useful, but are expressed differently from the absolute risks discussed here.

that, under certain DES exposure conditions, there occurs a greater number of those types of cancers than the number that occur in the absence of DES exposure. The risk assessor also means that DES has contributed to the cause of those extra cancers, in the sense that its presence directly brought about certain changes (probably related to its endocrine tissue-stimulating properties, a property of all estrogens) that enhance tumor development. A goal of risk assessment is to estimate the extra risk caused by a toxic or carcinogenic chemical over that which exists with no exposure to the chemical.

This notion of risk is, as described in Chapter 6, not to be confused with what is called a *risk factor*. Physicians say that people who are overweight are at increased risk of heart disease. But this does not necessarily mean that the heart disease is caused by the obesity. Rather, obesity is what is called a *risk factor*: physicians know from much correlational data that heart attacks occur more frequently in individuals who are overweight, but do not have compelling evidence that it is the extra weight that is the direct cause of those heart attacks. Other factors, correlated with both obesity and heart attack rates, are probably the underlying cause. Knowing risk factors is exceedingly important, because physicians can use this information in treating disease. But it is important to keep in mind the difference between a risk factor and risk, which carries with it the connotation of a true contribution to causation.

There is a broad commonality of approach among the federal agencies that engage in risk assessment (EPA, FDA, OSHA, CPSC, USDA), and those that do so for more general public health purposes or to advise regulatory agencies (the ATSDR, NIOSH), their counterparts in state agencies and in the European Union, and the various scientific institutions and committees of the WHO and FAO and our own National Academies. Some differences in approach among these various bodies have appeared from time-to-time, and this is troublesome, but it does not serve the purpose of this book to dwell on such aberrations.

As what we have termed the regulatory approach is discussed, it should be kept in mind that much of what is done in the conduct of risk assessment is to ensure a high degree of consistency and predictability in the face of scientific uncertainty. There are many scientists who will disagree with the regulatory approach. Sometimes the disagreement arises in specific cases, where evidence has arisen that seems to question one or more of the typical regulatory defaults. Other scientists may find the whole process, because it is dependent upon untested (and maybe some currently untestable) default assumptions, scientifically suspect, and to be avoided altogether. (Although I have found that most scientists in the latter category, when made to appreciate the decision-making contexts under which risk assessments are conducted and the unacceptable consequences of avoiding the questions risk assessments are designed to answer, come to accept the importance of having a scientifically incomplete, but nevertheless systematic risk assessment procedure in place.)

## Organizing for risk assessment

Defining the risk assessment problem to be evaluated should precede entering the four-step process set out in Figure 7.1, Chapter 7. This means identifying the population that is to be the subject of the assessment, and specifying the conditions under which it is or may come to be exposed to a chemical or mixture of chemicals. Formulations of the problem might be similar to any of the five examples offered at the beginning of Chapter 7.

For the purposes of risk assessment the exposed individuals are, in a way, *hypothetical*, not actual people. By this is meant that they will be assumed to exhibit certain characteristics that make it possible to reach general conclusions regarding the magnitude and duration of their exposure to the chemical of interest, and also their relative sensitivity to its toxic effects. It may be that there are actual people in the population having characteristics closely resembling those assumed by the risk assessor, but it is not possible to know (except in highly unusual circumstances) who those people are.[2]

Some risk assessors describe the process of setting up for risk assessment as developing a scenario. A scenario is a description of the population that is of interest and the way such a population is or could become exposed to a chemical or group of chemicals. Some typical scenarios for risk assessment are set out in Table 8.1, in abbreviated form.

The development of scenarios to be explored often occurs within a regulatory or public health context, in which institutions such as the EPA, the OSHA, or the NIOSH are attempting to fulfill their

---

[2] In Chapter 10 we deal with the problem of disease or toxic injury causation in actual individuals; the regulatory approach described in this chapter is not especially helpful when we are faced with this type of question.

Table 8.1 *Some scenarios for risk assessment*

| Scenario | Population | Source of chemical(s) | Chemical | Pathway to individuals in population | Routes of exposure |
|---|---|---|---|---|---|
| A | Individuals residing near hazardous waste site | Hazardous waste site | Volatile chlorinated solvents | Ground water used for drinking, bathing, cooking | Ingestion, inhalation, dermal contact |
| B | Children living in 1950s housing | Paint | Lead | House dusts containing paint particulates | Inhalation, ingestion of dust |
| C | Foundry workers | Foundry sands | Silica crystals | Suspended dusts | Inhalation |
| D | Consumers of beverages in polyvinyl chloride (PVC) containers | Container | Vinyl chloride (residual monomer present in PVC) | Migration of vinyl chloride to beverage | Ingestion |
| E | Consumers of animal-based foods | Incineration emissions | Dioxins | Uptake into animals through contaminated feed, storage in animal tissues | Ingestion |
| F | Consumers of corn | Pesticide application | Atrazine | Residues of atrazine in corn products | Ingestion |
| G | Patients requiring cholesterol-lowering drugs | Prescription drug | One or more statin drugs | Manufactured capsule | Ingestion |

responsibilities to ensure that citizens are not denied the right to a non-harmful environment or work place. Research institutions may similarly explore such scenarios; and risk assessments based on these scenarios may be a form of publishable research. Manufacturers wishing to move certain products into the marketplace, and to keep them there, will have to conduct risk assessment studies (studies of scenarios D, F, and G, Table 8.1, are typically sponsored by private parties). Certain activist groups will find ways to pursue investigations of scenarios they believe are being neglected by the authorities.

In some cases investigators may come to decide that a particular scenario is subject to direct epidemiological investigation. Thus, for example, it may be feasible to identify specific subpopulations of foundry workers, or patients taking statins, that are good candidates for study, and estimates of risk can be obtained by direct examination of any elevated rates of certain medical conditions in those subpopulations. Although such direct epidemiological studies avoid the need for the types of inferences and extrapolations used in the typical risk assessment, they do not usually bring clear-cut answers regarding causation, and several such investigations may be necessary before the risk questions are satisfactorily answered (Chapter 6).

For several reasons, risk assessments of the indirect type – involving the four steps described in the preceding chapter, and having both scientific and science policy elements – are needed to provide useful answers for public health and regulatory decision-making. Often, the public health question is urgent and cannot wait for the relatively long time necessary to mount, conduct, and interpret one or more epidemiology studies. And there are many circumstances in which it is simply not feasible to conduct a useful epidemiology study: obtaining reliable information on either population health status or exposure becomes, as discussed in Chapter 6, extraordinarily difficult. As we shall see in the risk management chapter to come, many laws require that data be developed to demonstrate the safety of certain classes of products (food additives, pesticides, pharmaceuticals) prior to their marketing, and this requires the use of data from animal studies in risk assessment.

The case of pharmaceuticals is distinguished from those of food additives and pesticides, in that, after a series of animal studies to evaluate toxicity, these drugs are investigated through intentional human dosing studies, called clinical trials. We shall devote a separate section to pharmaceutical risk assessment at the end of the chapter.

## Getting started

Once a scenario is defined and the purpose of the assessment understood, the risk assessment framework, Figure 7.1, is the guide to the application of the scientific principles and information described in this book regarding toxicity, dose–response relationships, and human exposure assessment. In addition, as we have taken pains to point out in the previous chapter, default assumptions are needed to deal with gaps in scientific understanding. For our purposes we shall invoke the defaults preferred and explicitly specified by the EPA, while recognizing that similar though not necessarily identical defaults have been adopted by other institutions carrying out risk assessments. Some of the defaults pertain to the toxicology findings, some to dose–response analysis, and some to the human exposure assessment step. As we move through these steps we shall describe relevant defaults and show how they are typically used in the risk assessment process.

At every step of the way an attempt is made to present a typical approach, and the usual default assumptions; it must be recognized that individual assessments often contain (usually minor) deviations from what is presented here, but what is presented should capture the most important aspects of current chemical risk assessment practice.

Information for risk assessment derives from studies published in the scientific literature and also developed in private and government laboratories. Once the scenario to be investigated is defined, it becomes possible to identify the pertinent scientific literature, and to retrieve it for study and critical evaluation.

## Moving through the steps of risk assessment

What adverse health effects are related to exposure to the substance of interest, and how certain is our knowledge? This is the question to be pursued under Step 1, hazard identification. Epidemiological and experimental toxicology data are the principal sources of the answer to this question. Depending upon the substance under review, the depth and completeness of the epidemiological and experimental investigation will vary. At one extreme there will be no epidemiological information available, and animal toxicology data will be limited to one or a few short-term studies. If this is the case, the only risk that can be evaluated is that associated with relatively short-term human exposures. Even such a minimal evaluation cannot be performed it if is found that the available data are of such poor quality as to be uninterpretable. At this extreme, risk assessment must be forsaken, and we shall have to go in search of the means to mount laboratory or epidemiological studies.

For some substances, the available data, both epidemiological and experimental, will be substantial, and enough of it will be of sufficiently high quality to be useful for a rigorous hazard evaluation. Generally, high quality epidemiological evidence, adequate to establish causal relationships between a chemical exposure and one or more adverse health effects will, when available, take center stage in a hazard identification (the criteria for establishing causality are those discussed in Chapter 6).

When epidemiological evidence is limited, and insufficient to establish causation, it may remain important in the hazard identification step if it is supported by reliable animal data. Within the regulatory context, convincing animal evidence of toxicity, even in the absence of strong epidemiological evidence, or indeed any epidemiological evidence, will still be used for hazard characterization.

This use of animal evidence is based, in part, upon its scientific standing, but it is also based upon a "science policy" decision – it is one of the "defaults" present in the risk assessment process. Even in the absence of specific knowledge that the response detected in a toxicology study is relevant to humans, it will be assumed to be so – unless other data arrive to demonstrate that it is *not* relevant to humans (see below what is meant by "other data"). Regulators and public health policies generally call for action even when the evidence regarding adverse health effects does not rise to the level necessary to establish causation in humans.

This brief sketch of the types of considerations that comprise the hazard identification step should provide some idea of what is sought. It can be said to include a "weight-of-evidence" evaluation of the available data to identify the full range of adverse health effects that can be related to the substance of interest, and to provide a clear picture of the quality and quantity of the available evidence. Even for a single substance the evidence supporting some of its associated adverse effects will be stronger than that supporting others. This difference is often a function of dose and the fact that it is easier to establish cause–effect relationships when effects (usually high-dose) are very severe and uncommon, than when they are subtle and common.

The hazard identification step should include an investigation into any scientific literature, if it exists, pertaining to the biological mechanisms underlying the substance's toxicity. This may provide information regarding, for example, the relevance of animal toxicity findings to humans, and the relevance of high-dose findings to responses that may occur at low-dose. If the toxicity data have been collected by one route of exposure only, it becomes important to determine whether there are data available to allow inferences to be made regarding possible effects associated with other routes of exposure. A third chapter on risk assessment (Chapter 9) will offer some specific examples of how these types of data can be used to support departures from the usual default assumptions.

A full hazard identification can thus be a large undertaking, drawing upon the collective skills of epidemiologists and toxicologists, and even pathologists, medical experts, and experts in mechanistic toxicology. It is typically presented as a narrative statement, and should set the stage for the remaining steps of the risk assessment. A typical hazard identification might conclude, after a comprehensive review and evaluation of all the relevant evidence, with statements of the following type (much abbreviated from what they should be).

- The evidence from several epidemiological studies of occupational cohorts and experimental studies in animals is sufficient to establish a causal relationship between inhalation exposure to chromium (+6) and cancer of the lung in humans.[3] Understanding of the mechanisms of cancer induction is relatively poor. Inhalation exposure to chromium (+6) is likely to pose a lung cancer hazard to humans. There is no evidence to suggest a cancer hazard if chromium (+6) is ingested, although studies on this question are limited.
- Fumes from fully volatilized gasoline (a complex mixture of hydrocarbons) have been shown to cause excess rates of kidney tumors in male rats. There is no such response in female rats or in either sex of mice. The epidemiological evidence regarding exposure to gasoline fumes and cancer in humans is equivocal. Mechanistic studies strongly suggest that the response is limited to male rats. The evidence that gasoline fumes pose a cancer hazard to humans is highly limited.
- Several occupational studies reveal that cadmium can slowly accumulate in kidneys and impair kidney function. These studies involved exposure by

---

[3] Chromium (+6) refers to one of several known valence (or oxidation) states of this metal. In the other most common oxidation state (+3) chromium is an essential element, and is not carcinogenic.

inhalation, but animal studies have shown that ingested cadmium can cause similar pathological changes. It is expected that kidney damage is a well-established effect of exposure to cadmium by either route. The mechanism of cadmium-induced kidney damage is well understood, as are the pharmacokinetic relationships between inhaled and ingested cadmium doses and the level accumulating in kidneys. It is possible to use this information to quantitatively relate the lowest level of cadmium in kidneys associated with toxic damage, and the external exposure.

Note that these narratives about hazards refer, in each case, to only one of several manifestations of toxicity, and that a thorough statement would have to supply similar information regarding the evidence for every type of toxicity that has been related to the chemical. Often, the information is organized by exposure duration; duration of exposure may affect toxic outcome.

Once the principal toxic hazards of a substance have been identified the assessors hope to find reliable and extensive information on the quantitative relationships between the magnitude of dose of the substance and the resulting hazards (toxic responses). Here we are at Step 2 of the risk assessment framework. For each hazard we would hope that the research studies that led to their identification have also produced the types of dose–response data described earlier, in Chapters 3, 4, and 6. Without such quantitative data development of quantitative risk assessments is not possible. Fortunately epidemiologists and toxicologists have done much in the past several decades to improve the quality of dose–response information, although the full development of quantitative dose data in the conduct of epidemiological investigations remains problematic. In some cases risk assessors, faced with using epidemiological data that lack quantitative exposure and dose information, and animal data that include highly reliable dose–response information, will be forced to choose the latter as the primary basis for risk assessment. Epidemiology data have had a somewhat less important role in risk assessment than have animal toxicology data, and this is likely to continue until methods for collecting quantitatively reliable exposure information in study populations are improved.

As soon as those responsible for identifying the critical hazards and their associated dose–response data have completed their work, it is time for some close collaboration with those who have been giving attention to the human exposure assessment, Step 3. This step is devoted to an evaluation of the exposure and consequent dose incurred by the populations of interest, and the routes (inhalation, ingestion, skin contact) by which they are or will become exposed. Generally, human exposure and resulting dose and risk are estimated for those members of the population experiencing the highest intensity and rate of contact with the chemical, although other, less exposed subgroups, and people experiencing average exposures, will frequently be included. Only when the human exposure assessment is well understood can the relationship between the available dose–response information and the human dose (or dose range) of interest for risk assessment be understood. We described in Chapters 2 and 3 some details of the human exposure assessment process, so simply describe here the types of information such assessments typically reveal, and some implications for risk assessment.

- Exposures in the population of interest will generally reveal that incurred dose is only a small fraction, and sometimes a very tiny fraction, of that at which toxic responses has been or can be directly measured, in either epidemiology or animal studies. Occupational populations (Table 8.1, Scenario C) may be exposed at doses close to those for which data are available, but general population exposures are usually much smaller. Thus, to estimate risk it will be necessary to incorporate some form of extrapolation from the available dose–response data to estimate toxic response (risk) in the range of doses expected to be incurred by the population that is the subject of the risk assessment.
- In some cases, the duration of exposure that is or might be experienced by the population of interest might not match that involved in the study. So, for example, dose–response information from relatively short-term exposures might in some cases be the only available information when the concern is long-term, or even lifetime exposure in the population that is the subject of the risk assessment. If the risk assessment is to be completed before new, long-term data can be developed, some justification will have to be found for extrapolation of the short-term data to estimate the consequences of long-term exposure.
- In some cases the available dose–response data will have originated from studies by one route of exposure (say, inhalation), but the population of interest is or might be exposed by other routes, say the oral one. Completion of a risk assessment prior to the development of new oral data will require a biologically justifiable method for extrapolating results from one route of exposure to another.

These three commonly encountered problems in dealing with the dose–response step of the risk assessment process (and there are others as well) are respectively referred to as the problems of: (1) high-to-low dose extrapolation; (2) extrapolation across exposure durations;

and (3) inter-route extrapolation. All involve extrapolation from a given set of dose–response data to estimate responses under different dose conditions. In some cases data will be available to guide at least some aspects of the extrapolation, but in most cases general default assumptions will have to be involved.

These types of extrapolation problems are not the only ones that have to be tackled. Here are some more:

- Often several different toxic hazards (neurotoxicity, organ toxicity, developmental toxicity, for example) associated with the substance that is the subject of the risk assessment, each with its own dose–response characteristics, will emerge from the first two steps of the risk assessment. Which of these should become the principal basis for the final risk assessment?
- If dose–response data from an animal study are selected as the principal basis for assessing risk, how are they to be applied (extrapolated) to the human population that is the subject of the risk assessment? This is the problem of interspecies extrapolation.
- Are all the humans who are members of the population under assessment expected to respond in exactly the same way to a given dose of the substance of concern? If not, how is *variability* in response among members of the human population to be considered?
- Perhaps there are some members of the human population under assessment (a subpopulation, such as children, women of child-bearing age, people with compromised immune systems, etc.) whose response to a given dose falls well outside the normal range of population variability. How can their risks be assessed?

These are the most important types of problem that arise which must be dealt with if risk assessments are to be completed. Again, there may be data available for some chemicals that allow reasonably accurate scientific answers for some of these questions, but as we emphasized in Chapter 7, scientific answers will generally be found wanting. Hence invocation of science policies – defaults. In Table 8.2, we find the most important regulatory defaults for risk assessment.

A similar set of defaults could be described for the human exposure assessment step. As noted, the regulatory approach tends to target those members of the population who are at the "high end" of exposures, and in some cases regulators, and other risk assessors, engage in something close to what is known as a "worst-case" exposure analysis. Here are three simple examples.

A petroleum refinery emits benzene. Assume a hypothetical resident lives at the plant's fenceline for a full lifetime and spends most of every

Table 8.2 *Typical regulatory defaults*

1. In general, data from studies in humans are preferred to animal data for purposes of hazard identification and dose–response assessment.
2. In the absence of human data, or when the available human data are insufficiently quantitative, or are insufficiently sensitive to rule out risks, animal data will be used for hazard identification and dose–response assessment.
3. In the absence of information to demonstrate that such a selection is incorrect, data from the animal species, strain, and sex showing the greatest sensitivity to a chemical's toxic properties will be selected as the basis for human risk assessment.
4. Animal toxicity data collected by the same route of exposure as that experienced by humans are preferred for risk assessment, but if the toxic effect is a systemic one, then data from other routes can be used, with appropriate adjustments.
5. For all toxic effects other than carcinogenicity, a threshold in the dose–response curve is assumed. The lowest NOAEL from all available studies is assumed to be the approximate threshold for the groups of subjects (humans or animals) in which toxicity data were collected. Alternatively, a benchmark dose (BMD) may be estimated from the observed dose–response curve, and used as the point-of-departure for risk assessment (see below and Box).
6. The threshold for the human population is *estimated* by dividing the NOAEL (or, alternatively, the BMD), by an *uncertainty factor* (UF), the size of which depends upon the nature and quality of the toxicity data and the characteristics of the human population. (The estimated human threshold dose has several different names, depending upon the regulatory context, see later.)
7. For carcinogens a linear, no-threshold dose–response model (Figure 8.1) is assumed to apply at low dose, unless data are available in specific cases to demonstrate that such a model is inappropriate.

day at home. Calculate the daily benzene dose this hypothetical person receives, and then estimate his cancer risk.

Assume a person consumes a particular food at a very high rate (90th percentile of consumers of that food), and that every mouthful for a whole lifetime contains a pesticide residue at the maximum allowed concentration. Calculate this person's dose and risk.

Assume a worker is exposed to the maximum allowable air concentration of a work place carcinogen, eight hours every day, five days every week, for a working lifetime of 40 years. Calculate the worker's dose and resulting risk.

Just what do risk estimates based on such assumptions mean? To varying degrees they are all fairly improbable, "worst-case" exposure scenarios. They are, however, easy to work with – doses can be estimated under these assumptions with very little real information and the calculations can be done by a high school chemistry student. Additionally, if risks estimated under such assumptions are very low, we can be doubly assured that actual risks are probably not significant. Of course, if they are high, as they often are, one wonders what to make of them. The best course would be to return to the drawing board and to try to obtain some real data to substitute for the assumptions used. In any case, risk estimates developed under such assumptions, it should always be emphasized, probably apply to few real people.

Regulatory agencies also attempt to develop more realistic estimates, but this is difficult, and a scientific consensus on just what exposure pattern should be presumed desirable for risk assessment is not available, except for a few circumstances (food additives, human drugs, and a few others). Much attention is now focused on methods to develop information on the full distribution of exposures in a population, but this can be technically difficult to achieve.

We can now proceed to demonstrate how scientific information and the regulatory defaults of Table 8.2 can be applied. It is useful and important to separate the dose–response evaluations into those used for substances that produce their toxic effects through threshold mechanisms, as these terms were described or used in Chapters 3 and 6, and those that may involve no-threshold mechanisms. As a practical matter, only carcinogens have, to date, been treated as belonging in the latter category.

## Substances acting through threshold mechanisms

Having begun this chapter with a primer on risk, and having devoted the previous chapter and this one to the general topic of risk assessment, it may seem odd to reveal at this stage that the current approach to substances producing their toxic affects through threshold mechanisms does *not* result in an explicit, quantitative statement of the *risks* they may pose. Rather, the toxic "risks" of these substances are described by extrapolating from the observed dose–response data to estimate what might be called a "threshold dose" for humans, and comparing that dose ($D_T$) to the "high-end" dose incurred by the human population under assessment ($D_H$), as that is estimated in the human exposure assessment step. Risk characterization (Step 4), in its quantitative aspect, involves a simple calculation of what is sometimes called (unaccountably) a hazard index (HI).

$$HI = D_H/D_T$$

If HI is less than one, it may be inferred that even those in the human population at the high end of exposure do not incur doses in excess of the estimated threshold dose, and so are unlikely to be at risk of toxicity. For HIs that exceed one, it may be inferred that some fraction of the population is at risk of toxicity, and the larger the HI, the larger the (non-quantified) risk.

After all the build-up to the subject of risk assessment, this rather crude approach must seem a little regressive. It is, in fact, an approach that has been in use for more than half a century, and that has been updated and refined only in the ways in which we think about what we are up to when we derive a $D_T$.

In the early 1950s, two FDA scientists – Arnold Lehman and O. Garth Fitzhugh – were exploring methods for setting protective levels for human intake of food chemicals, based on animal toxicity data. These scientists reviewed the rather scant literature available at that time on the relative sensitivities of humans and test animals, and the degree of variation in sensitivity within the human population. Based on a reading of the available literature, they concluded that a "safety factor" of 10 for each of these two variations – relative sensitivities of humans and rodents, and variation in sensitivities among humans – would be adequately protective. So they proposed that what they termed an acceptable (or allowable) daily intake (ADI) for chronic human exposure to chemicals, as might occur, for example, for food additives or pesticide residues in food, should be set at 1/100th of the chronic rodent NOAEL.

The Lehman–Fitzhugh approach has been very widely used for setting limits on exposures to chemicals, not only in food, but in all other environmental media, but it has undergone significant refinement in recent years. The EPA, and others, for example, now uses the term "uncertainty factor" for those factors that reflect a true scientific uncertainty, and distinguishes these from "safety factors," which reflect the injection of policy judgments that go beyond scientific uncertainties in the establishment of "acceptable" intakes. The EPA has dropped the term ADI, and instead calls the derived human threshold estimate an RfD – toxicity reference dose – removing the inference of "acceptability," which, they say, carries with it the connotation of a non-scientific, value judgment. The ADI, the RfD, and other, similar terms are said to provide an estimate of what, for convenience, we have termed an approximate threshold dose $(D_T)$ for humans. How is the $D_T$ derived?

To do this we need to rely upon the regulatory defaults and assumptions listed in Table 8.2, and proceed as follows:

(1) The toxic effect occurring at the lowest dose (the most sensitive indicator of a chemical's toxicity) is selected as the critical health concern for risk assessment.

(2) The NOAEL for that effect is identified.

(3) If the NOAEL is from a study of less-than-lifetime duration and it is to be used to estimate a $D_T$ for lifetime exposure, the NOAEL is divided by what is called an uncertainty factor (UF) of 10, because it is reasonably clear that toxic effects become apparent at lower dose as the duration of dosing increases. A UF of 10 is usually regarded as sufficient for converting a NOAEL based on exposure over a fraction of a lifetime to estimate a lifetime NOAEL. The factor of 10 is supported by some empirical evidence from observations on chemicals for which both subchronic and chronic data are available (the chronic NOAEL is rarely less than one-tenth of the subchronic NOAEL).

(4) If the NOAEL is derived from a study of human populations, it is divided by an uncertainty factor that is meant to account for the extent of variability in toxic response in the human population that is the subject of the risk assessment. Uncertainty factors typically take values from 1 to 10, and the value selected depends upon the characteristics of the population that was studied to derive the NOAEL, and their relation to those of the population that is the subject of the risk assessment. A population that is relatively homogeneous in nature (e.g., a group of workers who are generally adults of early-to-late middle age and, on average, healthier than the general population) is likely to display a narrower range of variability than one that is highly diverse (the one we have referred to as the general population). Thus, a NOAEL derived from a study of workers may be applied to another population of workers without any adjustment (a UF of 1), or only a small one, whereas applying the same NOAEL to the general population usually signals the need for a larger UF. A UF of 10 is common for the latter.

(5) The UF of 10 for human variability is not intended to reflect the full range of variability in the population. Rather, it reflects the difference in response between a hypothetical "average" member of the population and those at the "high end" of susceptibility. Thus the full range of variability – that covering the difference between "most" and "least"

susceptible – could be a factor of 100. The NOAEL from a human study is thus assumed to apply to the hypothetical "average human."

(6) If the NOAEL is derived from an animal study, a UF of 10 is used typically to extrapolate to the "average" human. This UF has some limited empirical basis. The available data suggest that most animal–human differences in response are less than a factor of 10, but the evidence supporting this conclusion is not strong.

(7) Additional factors may be applied if there are significant shortcomings in the available data that are a cause for concern. If, for example, data relating to the effects of a chemical during the developmental phase are absent, and there are reasons to suspect that the chemical could have such effects (it may, for example, be structurally related to a known developmental toxicant) an additional factor may be applied to the NOAEL.

Suppose there are several animal studies available for a chemical, including some which have involved long-term exposures, and some relating to possible reproductive and developmental effects. Assume data for both mice and rats are available, and that there are well-defined NOAELs for each of the identified toxicity endpoints. The lowest NOAEL from all those available derives from a chronic study, and is 10 mg/(kg day).[4] Given such a data base, and the defaults listed above, a $D_T$ for this chemical applicable to the general population, would be derived as follows

$$D_T = \frac{\text{NOAEL}}{\text{UF}} = \frac{10 \text{ mg/(kg day)}}{10 \times 10} = 0.1 \text{ mg/(kg day)}$$

Here a UF of 10 covers extrapolation from animals to "average" humans, and another UF of 10 covers human variability in response (the difference between "average" and "high-end" susceptibility). No other types of extrapolation are needed for a substance having such a robust data base.

In some cases chronic toxicity data may not be available. Thus, if the NOAEL were derived from a subchronic (animal) study, an additional UF of 10 would be applied ($D_T = \text{NOAEL} /10 \times 10 \times 10$).

Consider a case in which the chemical has been subjected to substantial epidemiological study, and adverse effects of the exposure have been identified, and a NOAEL has also been determined.

---

[4] In recent years a new procedure has been adopted by the EPA for dealing with responses in the range of what we have been calling the NOAEL. The procedure is called benchmark dosing, and the benchmark dose has now replaced the NOAEL in these types of evaluation. We shall continue to use the NOAEL in our discussions, but the relationship between a NOAEL and a benchmark doses is explained in the Box, and in the section to come on carcinogens.

Methylmercury, the fish contaminant that was described in Chapter 4, is in this category. The problem of choosing among the various studies of the effects on mental development of children born of mothers consuming large amounts of fish is difficult, because one such study (conducted in the Seychelle Islands) appears not to reveal any developmental difficulties, whereas another, conducted in the Faroe Islands has identified learning difficulties in similarly exposed children. There are several possible explanations for the differences observed between the two study populations, but these explanations are of peripheral importance to what we are attempting to explain here. Suffice it to say that the EPA has chosen to rely upon the Faroe data, and has identified a level of methylmercury intake that qualifies as a NOAEL. That intake corresponds to a dose of 1 mg/(kg day). Because the dose–response information derives from a study involving individuals thought to be close to the "most sensitive" members of the human population, the EPA did not use a full factor of 10 to deal with human variability, but instead used a factor of 3 (it was thought that the Faroe children were not likely to be 10 times less susceptible than the most susceptible children in the general population to which the $D_T$ would be intended to apply). The EPA did incorporate another factor of 3 to account for some perceived deficiency in the data base. A total factor of $3 \times 3 = 10$ (government arithmetic) was thus used to derive a $D_T$ (an RfD) for methylmercury of 0.1 mg/(kg day).

---

### Benchmark doses

In recent years the EPA has come to rely less upon observed NOAELs, and more on what are called benchmark doses (BMD). As noted in Chapter 3, observed NOAELs cannot be taken as accurate indicators of the true experimental threshold dose. Their values are influenced heavily by experimental design. The BMD concept was developed as a way to eliminate the influences of experimental design. This goal is accomplished by applying a simple curve-fitting procedure to the observed dose–response relationship and identifying the dose that corresponds to a 10% response rate. The actual experimental doses are not likely to produce a 10% response exactly, so extrapolation along the fitted dose–response curve will usually be necessary to find the dose producing a 10% response (see Figure 8.1). The extrapolation distance will usually be very small, so there is little uncertainty in this process. Benchmark doses estimated from different dose–response curves all represent 10%

response rates, so using the BMD as the point-of-departure for applying uncertainty factors (or for applying low-dose extrapolation to carcinogenic responses) ensures a greater degree of uniformity in the risk assessment process than does reliance upon experimental NOAELs. See Figure 8.1, and further discussion in the text on carcinogens.

Each of the choices we have described involves a combination of scientific data, science policy assumptions ("defaults") and some additional judgments (e.g., about the importance of various data base deficiencies). With the background just presented you should be able to understand any such derivation offered by the EPA or other regulatory agencies, with two exceptions.

First, the procedure now used by the EPA for inhalation data differs from what we have described above, in that the ten-fold factor for interspecies extrapolation (animal-to-human) is dropped in favor of a specific model that describes the well-known physiological differences between animals and humans that affect the relative rates of movement of a given administered dose of a chemical in the respiratory tracts of animals and humans. These physiological models provide fairly accurate predictions of the relative doses of chemicals delivered into the respiratory regions of animals and humans who have received identical administered (inhaled) doses. The estimate of "delivered dose" offers a well-accepted scientific approach to at least part of the problem of interspecies differences. Details of the delivered dose calculations are beyond the scope of this book (see references in *Sources and recommended reading*).

The second difference pertains to the derivation of $D_T$ to protect workers. Historically, it has been assumed that workers are normally healthier than the general population, and show less variability in response to chemical exposure. Worker protection standards are thus routinely less restrictive than those covering the general population. Uncertainty factors used to derive worker protective limits are normally smaller than those used for the general population.

The use of the symbol $D_T$ is strictly my own, used for convenience. The actual names and symbols used depend upon the institutions involved (Table 8.3). None of these institutions refers to the values they derive as true population thresholds, but they are thought to be protective of large diverse populations.

Table 8.3 *Some terms used in the United States to describe levels of daily chemical intake thought to pose insignificant risks to health*

The acute exposure guidelines discussed in Chapter 4 have a similar basis to the toxicity factors listed here.

| | Occupational exposure | | General population exposure | | |
| --- | --- | --- | --- | --- | --- |
| Term | Source | Status | Term | Source | Status |
| Permissible exposure level (PEL) | Occupational Safety and Health Administration | Official regulatory | Allowable daily intake (ADI)[a] | Food and Drug Administration | Official regulatory |
| Threshold limit value (TLV) | American Conference of Governmental Industrial Hygienists | No official standing widely used by industry | Toxicity reference dose (RfD)[b] | Environmental Protection Agency | Official regulatory |

[a] ADI is also used by WHO/FAO and many other countries for substances intentionally introduced into foods. The term tolerable daily intake (TDI) is often used for contaminants.
[b] The EPA reports toxicity reference concentrations (RfC) for inhaled substances; PEL and TLV are for inhalation exposures, and ADI and RfD are for oral exposures.
Note: The term minimum risk level (MRL) is used by the ATSDR.

Table 8.4 *Selected PELs and RfDs used in the United States. PELs are meant to apply for 8 hours/day and 5 days/week; RfDs (and RfCs) are applicable to continuous daily exposure over a full lifetime*

The PELs and RfDs in this table may not be based on the same toxicity endpoints.

| Chemical | Occupational PEL (mg/m$^3$ air) | General population RfD (mg/(kg day)) |
|---|---|---|
| Toluene | 750 | 0.2 |
| Benzene | 1.0 | |
| Mercury | 0.1 | 0.0003 |
| Acrylamide | 0.03 | 0.0002 |
| Lead | 0.05 | Blood levels in children not to exceed 10 µg/dl. |

Table 8.4 provides a few examples of these various protective limits. Recall that RfD is for non-carcinogenic effects only; benzene is regulated as a carcinogen (see later).

There have been attempts to develop explicit risk estimates for agents presumed to act through threshold mechanisms. Some investigators have proposed the use of models which assume that toxic responses and the thresholds for them follow a certain distribution over the population. The use of such models may reveal how the distribution of response shifts as dose shifts. Unfortunately most toxicology data are not reported in a form that allows ready use of such distribution models.

## Using RfDs, TDIs, etc.

As stated at the outset of the section on threshold agents, risk is characterized (Step 4) by deriving what is sometimes called a hazard index, the ratio of known or expected doses incurred by the human population ($D_H$) to the RfD, TDI, ADI, or MRL. A hazard index exceeding 1.0 suggests a risk; that is, it can be taken to mean that some members of the population are exposed at levels exceeding the estimated population threshold. It is an unquantified risk, in two senses. First, although the RfD (or other estimates of "safe" dose) is considered to be without significant risk, it is derived and expressed in a way that affords no hint of the probability that doses at or near it actually present risk, except to say the risks are likely to be small. And second, for incurred doses that exceed the RfD, we are given no insight into the magnitude of excess risk. Many analysts find that such an approach offers little guidance to decision-makers, because it keeps risk information hidden. But it is the system we have for all forms of toxicity except carcinogenicity, and its simplicity and lack of explicitness regarding risk make it relatively easy for risk managers to use and to explain – they can avoid the tough question of how much risk is actually being tolerated under any given decision. Many analysts refer to the HI approach, or whatever its equivalent may be called, as a "safety assessment," a comforting phrase.

If we are concerned with occupational exposure or inhalation exposure, we turn to PELs, TLVs, or RfCs. The first two are expressed as air concentration averaged over an 8-hour work day, while the last is meant to apply 24 hours a day, 7 days a week, for a full lifetime. Assessments of risk are made by comparing the air concentrations experienced by populations with the appropriate estimate of a population threshold, and again concerns about risk arise when these ratios exceed one (1). For no well-defined reason, regulators usually do not raise serious alarms until those ratios reach 10 or so; the RfD or TDI or RfC does not represent a sharp dividing line between "safe" and "unsafe" exposures, and its inherent imprecision leaves room for risk managers to exercise some degree of flexibility when it comes to considering reducing human exposures.

The RfDs and TDIs are often used to establish *regulatory standards*. Such standards usually specify a limit on the allowable concentration of a chemical in an environmental medium. The process is not difficult to understand. The RfD and its related estimates of population thresholds is a dose, typically expressed in mg/(kg b.w. day), that is considered to be without significant risk to human populations exposed daily, for a lifetime. Consider mercury, a metal for which an RfD of 0.0003 mg/(kg b.w. day) has been established by the EPA, based on certain forms of kidney toxicity observed in rats (Table 8.4). These are not the only toxic effects of mercury, but they are the ones seen at the lowest doses. Note also that we are dealing with inorganic mercury, not the methylated form that is neurotoxic.

Suppose a limit on mercury levels in drinking water needs to be set. The goal is to ensure that the RfD is not exceeded. To do this, the EPA first selects a hypothetical, average person, whose lifetime body weight averages 70 kg and who drinks an average two liters of water each day. For an RfD of 0.0003 mg/(kg b.w. day), the allowable daily mercury *intake* is:

$$0.0003 \, \text{mg/(kg b.w. day)} \times 70 \, \text{kg b.w.} = 0.021 \, \text{mg/day}$$

That is, a 70 kg person could take in 0.021 mg of mercury each day, and thereby receive a dose that does not exceed the RfD of 0.0003 mg/(kg b.w. day). If the 0.021 mg mercury were received entirely through drinking water, and two liters of water were consumed each day, then the "safe" drinking water concentration (with rounding) is:

$$0.021 \, \text{mg/(2l day)} = 0.01 \, \text{mg/l}$$

A drinking water concentration of mercury of 0.01 mg/l (10 ppb) gives rise to an intake of just under 0.021 mg of mercury each day, which corresponds to a dose of 0.0003 mg/(kg b.w. day).

Reference doses for pesticides are used in just this way to establish standards (called tolerances) for pesticide residues in food. Similarly, ADIs, derived for food additives, are used to establish allowable limits on amounts that can be added to foods. Data on rates of food intake are necessary for pesticides and food additives, in the way that water intake rates were used in the example given above.

## Carcinogens – no-threshold models

Figure 8.1 is the starting point for a look at carcinogens. The figure shows a typical dose–response curve for a carcinogen (dose versus lifetime risk of tumor development, as described in Chapter 6). The observed dose–response curve is investigated and the "best-fit" model is derived using standard curve-fitting procedures. The statistician estimates a statistical upper confidence bound on the fitted curve, and then identifies the dose that is estimated to produce a 10% response (lifetime cancer risk of 0.1). The latter is called the lower confidence limit on the benchmark dose (BMD).[5] The lower bound on the BMD is taken as the point-of-departure (POD) for low dose risk estimation. In essence, a straight line is drawn from the POD to the origin, and risks in the range of human exposures are simply derived from that

[5] Because it is derived from the upper confidence bound on risk, the BMD is actually the lower confidence bound on the dose corresponding to a 10% risk. Statistical confidence bounds are used to account for expected variability in observed data. Their use adds an element of additional caution to the extrapolation process. See later.

straight line (see Figure 8.1, Close-up). The use of the BMD in this fashion ensures that the point-of-departure for low dose extrapolations for all carcinogens corresponds to the same 10% response rate. The straight line down to zero – a linear, no-threshold model (LNT) – is meant to place an *upper bound* on the human risk in the range of human exposure. Note also that if the dose–response data are derived from animal studies, another factor, called an interspecies scaling factor, is used to "scale" from animals to humans. The scaling factor is discussed later, following a discussion of the basis for the procedure for carcinogens we have just sketched.

For the range of human doses illustrated in Figure 8.1 (Close-up) we would say that the upper bound on excess lifetime cancer risk lies in the range of 0.000 000 8 (8 in 10 million, or $8 \times 10^{-7}$) to 0.000 008 (8 in one million, or $8 \times 10^{-6}$). Actual risks are unknown, but are not likely to exceed these upper bound limits. "Excess lifetime cancer risk" means the risk incurred over a full lifetime above that incurred in the absence of exposure to the carcinogen.

Evolution of the notion that carcinogens, unlike other toxic agents, act through mechanisms for which there is no definable threshold was set out in Chapters 5 and 6. Recall that "no-threshold" should not be taken to mean that any dose greater than zero can "cause" cancer; rather, it means that there is some increased probability of cancer development at any dose greater than zero. This probability, or risk, is thought to increase as dose increases. Epidemiologists and those who conduct animal experiments on carcinogens tell us that only when risks become very large (say 10% or more for animal studies, or a risk doubling for epidemiology studies) are they able to actually measure those risks. As noted many times in this book, typical human population exposures to most chemical carcinogens present in the environment occur at levels far below those at which risks can be directly measured.

When in the 1960s and 1970s the need for low-dose extrapolation became apparent, a number of mathematically oriented carcinogenesis experts proposed various models to link the observed dose–response relationship to the range of doses humans might be receiving. Some of these models were based on what was then understood about the biology of cancer development, while others were based on statistical curve-fitting and attempts to develop the most statistically sound estimate of low dose risks.

A model that attracted much attention was one that attempted to account for the multistage hypothesis of carcinogenesis, discussed in

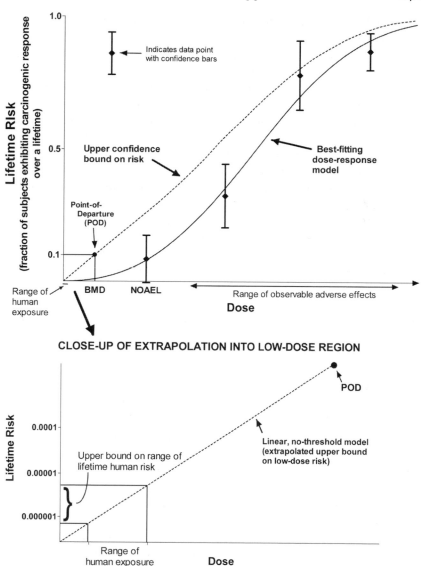

*Figure 8.1 Dose–response curves for carcinogens and illustration of low-dose extrapolation using linear, no-threshold model. Benchmark dose (BMD) is also illustrated.*

Chapter 5. During the 1970s several investigators developed a "linearized" form of the model, designed to place an upper bound on low-dose risk; this model yielded results regarding such risks not substantially different from those resulting from the currently used LNT.

The truth was that there was no established basis for predicting low-dose risks based on biological theories, and no basis for selecting one model over another. Moreover, models that provided the best "statistical fit" to the observed dose–response curve could not in any way be guaranteed to provide such a fit in the low-dose region. But it became clear that use of a simple procedure, based on an LNT extrapolation from a defined point on the upper confidence limit on the observed dose–response curve, as described in Figure 8.1, would be almost certain to yield what has come to be called an "upper bound" on the low-dose risk. The actual risk, which cannot be known, may equate to the estimated upper bound, but may actually fall on some other (unknown) dose–response curve that lies somewhere below the upper bound; it may even fall on a curve which shows zero risk in the low-dose region (if the carcinogen happens to follow a threshold pattern). It seems unlikely that the actual dose–response curve in the low-dose region somehow manages to rise above the straight line (Figure 8.1), and yield low-dose risk predictions greater than those predicted by the LNT.

So when this approach is used for carcinogens, it takes into account the no-threshold hypothesis (it predicts a risk at all exposures greater than zero), but there should be no pretense that we have arrived at an accurate prediction of risk. The LNT is the default used for carcinogens for low-dose extrapolation.

Three additional points need to be mentioned. First, if the observed cancer dose–response relationship derives from epidemiology data, the observed risks are relative, not absolute (the latter are usually reserved for data from animal experiments). Thus, for human carcinogens with reliable dose–response information (e.g., as exists for benzene, arsenic, chromium (+6), asbestos, and several other carcinogens), it is necessary to convert relative risks to absolute risks before extrapolating to low dose.

Second, there is a procedure for "scaling" doses between animals and humans, to take account of differences in body size and rates of various physiological processes. Interestingly, as the EPA and other regulators practice risk assessment, animal-to-human extrapolation for carcinogens is based on the use of such scaling factors, rather than the use of the ten-fold default factor used for non-carcinogens. This difference in approach has no known basis, at least to this author. Moreover, the carcinogen risk-assessment procedure does not include an uncertainty factor for variability in response among humans; again, the reason for its absence is not apparent, although it might be argued that the use of an upper statistical confidence bound, rather than the observed dose–response relationships, provides some accounting for variability.

Third, a so-far unmentioned default assumption concerns the measure of dose used in carcinogen risk assessment. There is some empirical basis for assuming that it is the *cumulative lifetime dose* of a carcinogen that determines risk. Thus, if a risk of *R* is related to a cumulative dose of 10 units of carcinogen, then any pattern of human exposure to the carcinogen that leads to a cumulative dose of 10 units will create a risk of *R*. Thus, a population incurring daily exposures that accumulate to 10 dose units in 10 years is at the same risk as a population incurring (much smaller) daily exposures that accumulate to 10 dose units only after a full lifetime, assumed to be 70 years. (The second population would be receiving only one-seventh the daily dose of the first.) Exposure assessment for carcinogens results in an estimate that represents cumulative lifetime dose, even if the dose is incurred over only a fraction of a lifetime. The scientific basis for cumulative dose as the determinant of cancer risk is not firmly established, so represents another default assumption. Indeed, many investigators have presented evidence that the pattern of exposure does indeed matter. Carcinogens that initiate the process may be much riskier if exposure to them occurs early in life, and those that promote may be much riskier if exposure occurs later in life. Regulatory risk assessments do not routinely account for these differences.

## Cancer slope factors

Under the regulatory risk assessment model, the result of applying the LNT to carcinogenicity data is an estimate of the upper bound on what may be called the "potency" of the carcinogen. By potency we refer to the *upper bound on lifetime cancer risk associated with one unit of average daily lifetime dose* (the usual expression of cumulative dose), obtained by extrapolation as shown in Figure 8.1. The carcinogenic potency is the *slope* of the straight line in Figure 8.1, at low dose. Slope

Table 8.5 *Selected potencies, or cancer slope factors, used by the US EPA.[a] For those based on animal data, the potencies include a factor for interspecies scaling (see text)*

| Chemical carcinogen | Oral exposure $(mg/(kg\ day))^{-1}$ | Inhalation exposure $(\mu g/m^3)^{-1}$ |
|---|---|---|
| Benzene | $1.5 \times 10^{-2}$ to $5.5 \times 10^{-2}$ | $7.8 \times 10^{-6}$ |
| Vinyl chloride | 1.4 | $8.8 \times 10^{-6}$ |
| Chromium (VI) | $-^{a}$ | $1.2 \times 10^{-2}$ |
| Methylene chloride | $7.5 \times 10^{-3}$ | $4.7 \times 10^{-7}$ |
| Cadmium | $-^{b}$ | $1.8 \times 10^{-3}$ |

[a] The slope factor expresses the upper bound on excess lifetime risk of cancer for one unit increase in dose; the latter is expressed as mg/(kg day) for oral exposure and as $\mu g/m^3$ for inhalation exposure. Risk is unitless, thus units for slope factors are $(dose)^{-1}$. See text for fuller explanation.
[b] Carcinogenic only when inhaled.

is the rise of that line – the increase in risk, shown on the vertical axis of the close-up in the figure, for each unit rise in dose, shown on the horizontal axis. The inherently more dangerous carcinogens exhibit greater extrapolated slopes than do the less dangerous ones. Some cancer slope factors are shown in Table 8.5.

The upper bound on lifetime risk can easily be estimated by multiplying the cancer potency by the number of dose units individuals are, or could be exposed to each day. This multiplication constitutes the quantitative component of risk characterization (Step 4) for carcinogens. That is, if potency has units of "upper bound on lifetime risk per unit of dose," and we multiply it by number of "dose units," the result is "upper bound on lifetime risk." This is the mathematical form of what we are doing when we are "reading" the risk directly from Figure 8.1.

The EPA's potency estimate for methylene chloride, shown in Table 8.5, is based on animal carcinogenicity data, and is 0.0075 in units of lifetime risk per dose of one mg/(kg b.w. day). (Human data regarding the carcinogenicity of methylene chloride are inconclusive.) One source of methylene chloride exposure is drinking water, where it can come to be present because of its wide use and escape into the general environment. Suppose the average methylene chloride concentration of a particular drinking water supply is 0.050 mg/l. If people drink 2 liters of this water each day, a total of 0.10 mg of methylene chloride will be consumed ($0.05 \times 2$). If these people average 70 kg body weight, then the average daily methylene chloride dose will be:

$$\frac{0.1\,\text{mg/day}}{70\,\text{kg}} = 0.0014\,\text{mg/(kg\,day)}.$$

If people receive this dose of methylene chloride each day for a full lifetime, then according to the EPA dose–response model, the *upper bound on excess lifetime cancer risk* is:

$$0.0075\,\text{per one mg/(kg\,day)} \times 0.0014\,\text{mg/(kg\,day)} = 0.000\,01(1 \times 10^{-5}).$$

If this risk is accurate, it means that 1 of every 100 000 people experiencing an average intake of methylene chloride through drinking water of 0.0014 mg/kg b.w., each day for a full lifetime, will develop cancer over a 70-year lifetime. This risk may also be expressed as 10 per million.

But remember, this risk (extra lifetime probability of cancer) is accurate only if all of the following hold:

(1) Methylene chloride is a human carcinogen.
(2) The animal carcinogenicity data provide an accurate picture of human response, in both the nature of the response (cancer) and its quantitative aspects (potency).
(3) The linear, no-threshold, dose–response model is accurate for very low exposures.
(4) People actually achieve the estimated level of ingestion every day for a lifetime.

The estimated risk will be *greater* than the actual human risk if one or more of the following is correct:

(1) Methylene chloride does not increase cancer risk in humans under any conditions (in which case actual risk will be zero).
(2) The animal model is more sensitive to methylene-induced carcinogenicity than are human beings.
(3) There is a threshold in the dose–response curve, or the curve drops toward zero risk more quickly than is suggested by the LNT.
(4) Exposure does not persist for a full lifetime or is otherwise less than that indicated.

The estimated risk will be *less* than the actual human risk if:

(1) Human beings are more sensitive to the carcinogenic effects of methylene chloride than are the experimental animals used for hazard and dose–response modeling.

(2) The actual dose–response curve falls toward zero risk more slowly than is indicated by the LNT (it is a so-called superlinear model).
(3) Exposures are actually higher for some people because they are consistently exposed to water concentrations that exceed the average level.

Most scientists would hold that these unknowns and uncertainties in the regulatory risk-assessment model would tend to favor risk overestimation rather than underestimation or accurate prediction. While this view seems correct, it must be admitted that there is no epidemiological method available to test the hypothesis of an extra lifetime cancer risk of about 10 per 1 000 000 from methylene chloride in drinking water. The same conclusion holds for most environmental carcinogens. It is also the case that more uncertainties attend the risk assessment process than we have indicated above.

Risks have so far been presented in quantitative terms, with a discussion of some of the conditions that would have to be true if the risk were to be considered accurate. Additional commentary on the likelihood that these conditions are correct, and the likely effect on risk (to increase or decrease it) were any not to be, is a critical part of the risk characterization, Step 4 in the risk assessment framework.

In most cases, the effect on estimated risks of alternative assumptions, and the probability that any such alternative is correct, cannot be estimated quantitatively. So the risk characterization includes both quantitative expressions and descriptive commentary. Too often only the quantitative expressions are given much weight, both by decisionmakers and the public. Numbers are easier to work with than is descriptive material, but this is no excuse for not trying to judge how close a particular estimated risk is likely to be to the true risk. The differences between risks estimated by the regulatory methods described and the true but unknowable risks will no doubt vary among chemicals. The data available for different chemicals are highly variable, and therefore so is the plausibility of the various science policy assumptions that are generically applied to all. The descriptive material accompanying the quantitative estimates should capture these important differences. Unfortunately many risk assessors have not yet learned to express these matters with both clarity and in a way that captures the underlying science and its limits.

## Actions based on hypothesized risks?

Many chemical risks such as those of methylene chloride in drinking water, are "calculated," not measured – that is, they are based not only on scientific data, but also on various sets of assumptions and extrapolation models that, while scientifically plausible (they fall within the bounds of acceptable biological theory), have not been subjected to empirical study and verification. Indeed, the results of most risk assessments – whether expressed as an estimate of extra cancer risk or an RfD – are scientific hypotheses that are not generally testable with any practicable epidemiological method. As we have said, there are no practical means to test whether methylene chloride residues in drinking water at 0.05 mg/l increase lifetime cancer risk in humans by 10 in 1 000 000, as hypothesized above. The tools of epidemiology are enormously strained, indeed, when called upon to detect the relatively low risks associated with most environmental chemicals. Without such a test, these calculated risks remain unverified.

Regulatory officials nevertheless act on the basis of such hypothetical risks ("hypothetical" definitely does *not* mean "imaginary"; it means that the risk estimates are based on certain scientific hypotheses and that they have not been empirically tested). Such actions are in part based on legal requirements (Chapter 11) and in part on the prudence that is a traditional feature of public health policies. The scientific information, assumptions, and extrapolation models upon which risk assessments are based are considered sufficiently revealing on the question of human risk to prompt risk-control measures. To put off such actions until it is seen whether the hypothesized risks are real – to wait for a human "body count" – is considered to be an unacceptable course.

The policy decision to act before science is certain does not, of course, dissolve the scientific uncertainties. Indeed, a strong argument can be made that an assessment of the type we have described should not pretend to represent normal science. Many of its outcomes are untestable with current methods; this alone might disqualify it as a true science.

The counter argument to the above rests on the premise that risk assessment outcomes, if not currently testable, might be in the future. Many scientific hypotheses are not testable at the time they are proposed; this does not mean they are "unscientific," assuming they are based upon and are consistent with all available knowledge. Moreover, risk assessors continue to urge the development of the type of data that will improve the reliability and testability of their predictions.

Some examples of developments in risk assessment that take advantage of mechanistic and other types of novel data and methods are the subject of the next chapter. This chapter also highlights some new and intriguing challenges for risk assessment.

## Pharmaceutical risk assessment

Pharmaceutical agents in many ways stand apart from the other classes of chemicals we have been discussing, in that extensive, direct study of their effects can be conducted in humans without breaching ethical standards. Once it is established through laboratory research that a compound has the potential to be a useful therapeutic, a formalized, widely accepted process of study and evaluation is undertaken to demonstrate whether it meets regulatory and medical criteria for safety and effectiveness. That process has evolved under the control of the FDA and its sister agencies in other countries, and those wishing to move a candidate compound into medical use are required to adhere to it. Clinical trials are the centerpiece of the process, in particular, the type of randomized, double-blinded, placebo-controlled trials discussed in Chapter 6. Evidence is collected during such trials of the therapeutic effectiveness of the candidate drug. The trials are conducted in three phases, with the first typically involving healthy subjects who have volunteered their time, usually for modest compensation. Information on possible pharmacokinetic interactions of the new drug with marketed drugs known to affect various P-450 enzymes, as described in Chapter 3, is often developed during the first phase. Successful completion of Phase One trials is necessary before trials involving patients are undertaken. Successful completion requires a convincing demonstration that the new drug does not cause excessive rates of significant side effects.

Data from pre-clinical studies in animals are used to select safe starting doses for Phase One trials. They also provide information for clinicians involved in trials on the targets for the new drug's toxicity, so that appropriate medical monitoring can be performed. Dose–response information and NOAELs from animal toxicity studies are thus primarily used to design clinical trials; data from preclinical studies may be included in the labeling of a drug once it is approved, but they are peripheral to ultimate assessment of a drug's risks. Risks are assessed based on the findings from clinical trials.

Cautious use of pre-clinical data, involving the insertion of safety (uncertainty) factors, provides the basis for selecting doses for the early clinical trials, but the predictive value of animal data is far from perfect, so that in most cases some adverse side effects will be observed in the human trials. Most often those effects are subjective in nature – headache, nausea, muscle pain – and could not have been suspected from animal studies. More serious side effects sometimes occur, and may put a halt to further development of the drug.

A widely used indicator of a drug's safety is the *therapeutic index* (TI). This index is usually expressed as the ratio of some measure of a drug's toxicity (a $TD_{50}$, for example) to a measure of the drug's expected efficacious dose (an $ED_{50}$, usually estimated from the dose–response curve for beneficial pharmacological effects). Large TIs predict greater safety. Whether such predictions stand can only be determined when clinical trials are completed.

All drugs will pose some degree of risk, and completed clinical trials are the primary source of information in this subject. Clinical trials do, of course, have limitations. The principal one concerns the ever-present problem of sample size. Rare side effects, if they exist, cannot generally be detected in clinical trials involving limited numbers of patients. Adverse drug reports and case-reports provide early clues to such effects; so-called pharmacoepidemiology studies may be mounted to evaluate such risks.

We shall review in Chapter 11 how this type of risk information is used in decisions about pharmaceutical use.

# 9

# Risk assessment III: new approaches, new problems

Scientists do not enjoy resorting to defaults, not only because of their uncertain scientific foundation, but also because their use takes some of the creative work out of the conduct of risk assessment. Uncovering novel approaches to the problems of any of the several forms of extrapolation used in risk assessment, or to the problem of variability in response among individuals, and identifying the types of scientific information that might be used to implement those approaches are highly challenging endeavors. If circumstances allow their implementation in specific cases, the accuracy of risk assessments might be improved. Basing regulations and public health decisions on more accurate risk assessment results should be welcomed by all.

There are several large impediments to achieving the goal of more accurate risk assessments. First, it often requires a considerable investment in the research necessary to uncover the types of information needed to replace default assumptions in specific cases. If one hypothesizes that di-(2-ethylhexyl)phthalate (DEHP, a real and important chemical) produces liver tumors in rodents by mechanisms that either do not apply to humans at all, or that do not operate at low (human) doses, or both, then there arises the question of what type of research information is necessary to test the validity of such hypotheses? If such research is actually carried out, then what type of results from that research would allow conclusions to be drawn about the validity of the hypotheses? In many specific cases creative and knowledgeable scientists can hypothesize alternatives to the usual defaults and ways to test their validity. But it often turns out to be difficult to arrive at a strong scientific consensus on the validity both of the alternative hypotheses and of the ways to test them.

Part of the difficulty involved in finding consensus stems from the fact that the regulatory approaches and the standard defaults rest in part on the insertion of a degree of caution in their selection. As we have noted in earlier chapters, this caution is sometimes said to represent a "conservative" approach to risk assessment, one designed, as far as possible, to ensure that risks are not underestimated. Most people would find such caution warranted, as long as it did not lead to extreme overestimates of risk ("hundreds exposed, thousands will die," as a friend of mine used to say). In any case, most alternative hypotheses about the risks of specific substances seem to suggest less public health risk than that predicted under the regulatory approach. Any such prediction is resisted by those who hold that any chance of risk underestimation is to be avoided. Hence, there is a heavy burden on those who would offer alternatives to achieve a high degree of scientific certainty, and to have their work scrutinized in excruciating detail. The EPA, for example, typically convenes scientific advisory boards to evaluate the quality and validity of risk assessments proposed on specific substances that are based on approaches resting on research not considered in the usual regulatory model.

A second impediment relates to the question of what we mean when we propose that we have achieved greater accuracy. What is usually meant is that we have relied upon information pertaining to the biological mechanisms that underly the production of toxicity, and have found ways to use that information to make inferences about risk. We will nevertheless have to admit that we have produced results that in most cases cannot be subjected to empirical test, so we cannot document greater accuracy in the usual way (e.g., by testing the results epidemiologically). Perhaps we should be claiming only that we have faith that greater reliance upon scientific understanding of underlying mechanisms should ensure that we are closer to the truth about risk. This is a tenet most scientists would adhere to, but achieving consensus in specific cases is, again, problematic.

Nevertheless, there are many scientists engaged in the adventure of finding ways to improve the conduct of risk assessment. This chapter is devoted to highlighting some of their achievements. The full basis for each of the many examples discussed is not presented, because this requires a far more complete explanation of, for example, pharmacokinetics or mechanisms of toxicity than we have offered in the earlier sections of this book. But a general account of the various approaches can still give a sense of the current direction of the practice of risk assessment.

Earlier discussions have shown that it is useful to think of two broad influences on the production of toxicity. First, the chemical has to make its way from the environment, into the body, and to the target site. The science of pharmacokinetics is devoted to understanding all the actions of the body on the chemical, including its conversion to both toxic and non-toxic metabolites and subsequent excretion from the body. Once at the target site(s), the chemical or, more usually, one or more of its metabolites, interacts with that site in some way that causes damage. The damage may progress or may be repaired, although continuation of dosing may limit the effectiveness of repair. The latter processes – interaction of the toxicant with the target site – gets us into the realm of pharmacodynamics.

## Roles for pharmacokinetic data

If it is possible to acquire an adequate understanding of pharmacokinetics, it may be possible in specific cases to document: (1) changing pharmacokinetic patterns with changing dose; or (2) pharmacokinetic patterns in the animal species used to develop toxicity data that are substantially different from those seen or expected in humans. Such differences would document that the usual defaults do not hold in those specific cases. The effects of using pharmacokinetic data in the risk assessment process instead of the usual defaults would depend upon what those data actually revealed.

A simple example might make this clearer. Suppose it were known that a 100 mg dose of chemical Z produced an extra 10% incidence of liver tumors in rats. Suppose further that we studied the pharmacokinetics of compound Z and discovered that, at the same 100 mg dose, 10 mg of the carcinogenic metabolite of Z was present in the liver. The usual regulatory default would instruct us to select the 100 mg dose as the point-of-departure for low dose extrapolation, and to draw a straight line to the origin, as in Figure 8.1. We are then further instructed to estimate the upper bound on risk at whatever dose humans are exposed to – let us say 1 mg. If the extra risk is 10% at 100 mg, then under the simple linear no-threshold model the extra risk at 1 mg should be $10\% \div 100 = 0.1\%$ (an extra risk of 1/1000).

Now let us assume that a clever pharmacokineticist arrived with some data showing that at the 1 mg (human) dose, only 0.001 mg of the carcinogenic metabolite could be found in the rodent liver. Recall that 10% (10 mg) of the metabolite was found at the administered dose of 100 mg. Now we see that only 0.1% (0.001 mg) of the administered dose reaches the liver, when that dose is 1 mg. If risk is governed by the amount of carcinogen in the liver, then the low dose risk should be much lower than that predicted by the linear model. Something happens to the pharmacokinetics of this compound between the administered dose of 100 mg and the dose of 1 mg, such that a much smaller proportion of the dangerous metabolite is produced at the low dose. Detoxification mechanisms are highly effective, in this circumstance, at low doses, and become less effective as dose increases. In technical terms, the enzymatic processes leading to detoxification are overwhelmed (they become "saturated") at some dose between 1 and 100 mg. If we could verify all of this, we should be able to adjust the risk downward by perhaps 100 times (10% of administered dose at the experimental dose, to only 0.1% of the administered dose at the level of human intake).

This idea is not complex, but acquiring the data to document it is not straightforward. Moreover, it can only be documented thoroughly in rodents because we cannot ordinarily perform the necessary pharmacokinetic studies in humans. So we would have to explain the basis for believing that the same *dose-dependent changes in pharmacokinetics* might occur in humans. Although it is not possible to conduct thorough kinetic studies in humans, it is becoming possible to piece together comparative information from test-tube studies of human and rodent cells, and information on comparative rodent and human physiology and biochemistry, to develop mathematical models of pharmacokinetic patterns. This type of physiologically based–pharmacokinetic (PB–PK) modeling has been pioneered by just a few, quantitatively oriented toxicologists – Melvin Andersen, Rory Conolly, Harvey Clewell – and references to some of the novel approaches they have developed are provided at the end of the book in *Sources and recommended reading*. It is exceedingly innovative work, and the rather crude example presented here is just one of several kinds of application to which such modeling is being dedicated.

The example presented demonstrates one type of outcome such pharmacokinetic data and modeling may reveal. For chemical X it could well show that a much greater fraction of its carcinogenic metabolite is created at low dose than high dose. In such a case the linear extrapolation would understate the low-dose risk. Pharmacokinetics is governed by physiological and enzymatic processes that may change in effectiveness as dose changes, and even as route of administration changes. If there is some way to understand these changes, the relevant data may be incorporated into risk assessments, and used to replace the standard default assumptions.

Studies of pharmacokinetics could, in some instances, reveal that substantially different metabolic pathways are operating in the animal species used to collect toxicity information than operate in humans. It is often difficult to know how to use this type of information unless it is clear that the differences pertain to the metabolite(s) causing toxicity, and not the metabolites that have little or no role; acquiring definitive data on this can often be problematic.

Two of the more interesting uses of pharmacokinetic data in risk assessment involve the neurotoxic agents lead and methylmercury (Chapter 4). In the case of lead, epidemiological studies have typically involved the development of quantitative relationships between levels of lead in the blood and adverse health effects. Other measures of lead in the body have also been used. Levels in blood are now very easy to measure, and they do carry the strong advantage that they integrate cumulative exposures from many possible sources (water, food, paint, soil, air, consumer products). Current public health targets for lead are expressed as blood concentrations, typically in µg/dL (Chapter 4).

If it is necessary to conduct a risk assessment relating to a specific source of lead (let us say a suspect public drinking water supply), the typical risk assessment would require the development of an exposure assessment relating to that specific source. This ordinary type of assessment would result in some estimate of the range of daily doses (µg/kg b.w.) that individuals using the water could incur. But that dose estimate has no direct utility: (1) there are no RfDs developed for specific sources, but rather the target for health protection is based on a blood lead level; and (2) the individuals consuming water are no doubt exposed to other sources of lead that contribute to health risk.

To deal with this problem the EPA invested in the development and validation of a pharmacokinetic model that is capable of relating intake of lead to blood level. The model also allows the risk assessor to develop blood level estimates that integrate all sources of exposure. Using this model, it becomes possible to determine whether a specific source, such as our suspect water supply, is leading to exposures in excess of the target for all sources combined (this assumes that other sources do not contain levels of lead greater than normal, background levels). In fact, the model permits identification of the distribution of exposures in a given population, and estimation of the fraction of the population that may be at risk. Like all regulatory risk assessment models, it is intended for use in estimating risks for populations assumed to have certain common characteristics, and is not intended for use in describing exposures in specific individuals.

Pharmacokinetics has played a crucial and somewhat unusual role in the assessment of health risks from methylmercury. Some of the epidemiology studies of this fish contaminant involved the measurement of mercury levels in the hair of pregnant women, and subsequent measurements of health outcomes in their offspring (Chapter 4). Various sets of pharmacokinetic data allowed estimation of the level of methylmercury intake through fish consumption (its only source) that gave rise to the measured levels in hair. In this way it was possible to identify the dose–response relationship in terms of intake, not hair level. Once the dose–response relationship was established in this way, the EPA was able to follow its usual procedure for establishing an RfD (which is 0.1 µg/(kg b.w. day)).

Pharmacokinetic studies play an essential and crucial role in the development of pharmaceuticals. Typically, for safe and effective use of drugs, it is essential to understand the relationship between levels in blood over time (which should represent the levels required for drug efficacy) and the level that may cause harmful side effects. Pharmacokinetic studies are conducted preclinically in animals and in the early phases of clinical trials (it is ethically acceptable to study pharmaceuticals in this way, while similar studies of other chemicals are generally not permitted except in very limited ways). The possibility of interactions between a drug and other drugs or even foods that may affect pharmacokinetic behavior is critical to investigate. If, for example, a new drug is excreted from the body by enzymatic conversion to a metabolite, and another drug is known to affect the activity of that enzyme, it is possible that the new drug, if co-administered with the second drug, may be excreted more rapidly than it should be (so that efficacy is lost), or less rapidly than it should be (in which case safety may be jeopardized). The effects of foods or other medicines on drug absorption rates could be similarly harmful.

Pharmacokinetic studies are so important in the pharmaceutical arena, that they play the central role in FDA approval of generic versions of drugs. A generic drug manufacturer has to demonstrate to the FDA's satisfaction a capability to manufacture the drug with appropriate levels of quality. But beyond that, the generic manufacturer need only demonstrate pharmacokinetic equivalence between the generic product and the originally patented brand (generic products can enter the market once patents expire). Pharmacokinetic equivalence (or bioeqivalence) is adequate to demonstrate safety and effectiveness, based on the assumption that the identical (generic) compound will be equal in effectiveness and safety to the original drug if it is present in the body at the same concentrations and for the same time period.

So pharmacokinetics plays highly important roles in risk assessment, and there is much more in sight. Particularly interesting are discoveries of inter-individual differences in metabolic "handling" of chemicals that are related to genetic differences. The development of human genome information is allowing toxicologists and pharmacologists to evaluate chemical and drug behavior in the body based on genetic differences; such "toxicogenomic" and "pharmacogenomic" studies should allow, for example, discrimination between high- and low-risk subpopulations.

## Roles for pharmacodynamic data

Some of the general biological phenomena underlying the production of toxic effects were outlined in Chapter 3, and, in connection with carcinogens, in Chapter 5. Many different types of interactions between toxic substances (again, most often, one or more of their metabolites) and components of cells were illustrated. It was also shown that toxicants might interfere in various ways with extracellular processes to cause harm. As the technology available to study toxic mechanisms has improved, it has become possible to take advantage of this knowledge in the conduct of risk assessments.

The case of chloroform-induced carcinogenicity is a useful way to introduce this topic. Chloroform ($CHCl_3$) has a long history of use as an industrial and laboratory solvent. It is also a by-product of the chlorination of drinking water, and there is widespread exposure to it from this source. This substance induces liver tumors in mice and kidney tumors in rats. In its first efforts to assess the risks the EPA used the usual LNT methodology. But continued research on the mechanisms of chloroform-induced carcinogenicity provided data that raised significant questions regarding the applicability of this methodology, and the EPA has now altered its approach and adopted a threshold model.

The EPA became convinced, nearly 20 years after the initial reports of animal carcinogenicity, that the tumors resulted indirectly from the significant toxicity to liver and kidney cells induced by chloroform metabolites (the highly toxic and reactive phosgene being one of them) at high doses. The cytotoxic effects resulted in regenerative hyperplasia, which in turn increased the risk of neoplasia. In the absence of cytotoxicity no increased risk of tumorigenicity is expected. This conclusion was supported by much empirical evidence, including the fact that tumors were induced only when chloroform was administered by stomach tube, and not when it was administered through the animals' drinking water. Under conditions of gavage dose, where the animals receive a large dose over a short period of time, significant cytotoxicity was produced; in the case of drinking water exposure at the same total daily dose (which was necessarily ingested in much smaller aliquots than was the gavage dose), no cytotoxicity and subsequent hyperplasia were observed. Drinking water exposure, similar to that occurring with humans, did not deliver doses large enough to cause cytotoxicity. The EPA also concluded that $CHCl_3$ was not a significant genotoxic risk, so that

a mutagenic mode of action via DNA reactivity is not a significant component of the chloroform carcinogenic process.

The EPA thus concluded that the dose–response curve for chloroform-induced kidney tumors followed a threshold-like (or what might be called a strongly non-linear pattern). Given this, the EPA rejected the LNT approach and decided that an RfD could be used as a basis for regulating chloroform. The EPA then derived a BMD from the kidney tumor data; this turned out to be 23 mg/(kg b.w. day). But instead of dividing this BMD by a UF, the agency focused on an RfD that had already been developed for chloroform, based on the need to protect against the risk of its toxic effects (including those considered to be the necessary precursors to its carcinogenic effects). This RfD is 0.01 mg/(kg b.w. day), a value separated from the BMD by a factor of 2300. This large margin, to which the EPA has given the name "margin of exposure," was considered adequate (probably more than adequate) to ensure the absence of a cancer risk at the RfD. To say this another way, if the EPA were to apply a UF to the BMD for kidney tumors to derive an RfD, it would be less than 2300. Thus, the existing RfD could be used to protect against all adverse effects of chloroform, including its carcinogenicity.

Although saccharin is no longer widely use as a non-caloric sweetening agent, it did enjoy a nearly 100-year run before evidence of its animal carcinogenicity appeared on the scene during the 1970s. As described in Chapter 6, saccharin reproducibly induces bladder tumors in the offspring of female rats dosed at dietary levels greater than 3%. Saccharin, because it displays very low toxicity, can be consumed by lab animals in very large amounts, corresponding to astoundingly large daily doses, without any shortening of their lifespans. The finding of bladder malignancies at these extraordinary doses could not be disputed, and for complex legal reasons and because alternative non-caloric sweeteners soon became available, saccharin's use declined precipitously (although congressional action prohibited the FDA from banning the use of saccharin).

At first, the public and scientific debate over saccharin might be described as a shouting match regarding the general reliability of animal bioassays. There was very little understanding of the mechanism of saccharin's carcinogenicity, and so little basis for an informed debate regarding the relevance of animal findings to humans, or to the much lower doses that humans would ingest through its use as a food and beverage additive (recall the "defaults" on these questions: the results of such testing are considered relevant until proven through research not to be).

Research on saccharin and bladder cancer took off after the initial shouting matches abated. First, the sweetener was found to be clearly devoid of genotoxicity. Then, over two decades, it became clear that this compound induced, only at very high doses, a series of physiological changes (for example, changes in urine acidity, density, and volume) that, together with the development of damage to the lining of the urinary bladder due to the presence of tiny, saccharin-induced crystals, bring about a tumorigenic process. Saccharin's tumorigenic effects appear to be limited strictly to the region of high doses that induce these physiological changes. This body of mechanistic knowledge, together with the absence of any signal from a number of epidemiological studies, points to a threshold mechanism.

In its ninth report on "known human carcinogens and reasonably anticipated human carcinogens," published in 2000, the NTP removed mention of saccharin (which had been listed in the first report, issued in 1980). The experts advising NTP reviewed a massive amount of mechanistic and epidemiological research and concluded that

... saccharin is not reasonably anticipated to be a human carcinogen under conditions of general usage as an artificial sweetener.

This conclusion by no means brings into question the general value of animal studies. Instead it tells us that the specific findings in the case of saccharin are almost certainly limited to the doses that cause the various precursor physiological changes in urine and in the lining of the urinary bladder, and that such changes are not remotely likely in humans consuming the relatively low doses associated with artificial sweetener use.

One other interesting example of how mechanistic information can be used to inform the conduct of risk assessment concerns a number of chemicals that are called *peroxisome-proliferating* compounds. Peroxisomes are tiny organelles present in the cytoplasm of liver and other cells. They contain systems for generation of hydrogen peroxide and for oxidation of fatty acids and are important in the normal processing of fats in the body. The presence of large amounts of fatty acids causes the numbers of peroxisomes to increase – they proliferate – a phenomenon not typical of other organelles. The process of peroxisome proliferation is simply an adaptation to changing cell physiology. But, beginning in the early 1970s and with increasing frequency since, a number of chemicals and drugs have been found to be capable of causing high rates of peroxisome proliferation, leading to the presence of increased numbers of these organelles in cells, and a consequent increase in the fraction of cell volume occupied by them. The chemicals capable of causing peroxisome proliferation are diverse in structure, and identification of common structural characteristics has not been entirely successful. Among the prominent peroxisome-proliferating substances are the cholesterol lowering drugs called fibrates, a substance called di-(2-ethylhexyl)phthalate (DEHP) that is added to plastics to increase flexibility, and a substance called PFOA (perfluorooctanoic acid) important in the manufacture of Teflon® and other fluorinated polymers. These substances (and several others) are not only peroxisome proliferators, but also cause liver tumor increases, at least in rodents. It seems that species differ widely in their responses to peroxisome-proliferating substances. Mice and rats are both highly responsive, while guinea pigs, monkeys, and humans are less responsive.

It now appears that a so-called *receptor* on the surface of cells is a key player in the series of events leading to peroxisome proliferation. Cell surfaces are loaded with all kinds of receptors – typically proteins that can interact with, or bind to, molecules of many types, both substances normal to the body and foreign chemicals. The binding can activate the receptor by altering its physical shape, and such an alteration can lead to the activation of genes within the cell that initiate changes within it. Many drugs are designed to interact with cell receptors and induce desired changes in cells. The particular receptor responsible for peroxisome proliferation is called the "peroxisome proliferator-activated receptor" (PPAR, now known to be one of several subtypes, and labelled PPAR-*alpha*, to distinguish it from other PPARs). Activation of PPAR-*alpha* is considered to be the first and initiating event in a series of complex biochemical processes leading to peroxisome proliferation and what is called "oxidative stress" within cells. High levels of very reactive oxidants within cells can lead to DNA damage. Cell proliferation can ensue. The DNA damage may persist, and, with cell proliferation, become dominant; preneoplastic lesions are produced and these sets of preneoplastic cells can further proliferate and create tumors.

Not all of the biochemical events in this complex pathway from PPAR-*alpha* activation to tumors are completely understood, but much is known. It seems that at least some peroxisome-proliferating chemicals that also produce tumors in rodent livers do so through this pathway. If it can be demonstrated that such a mechanism is at work, then it seems that the risk of tumorigenicity for such compounds would be limited to doses that are sufficient to activate PPAR-*alpha* sufficiently to initiate the dangerous cascade of events within the cell. Experts have developed a number of experimental criteria that should be met if a compound is to be put in this class of carcinogens. Study of PPAR-*alpha* activation as a route of carcinogensis is an extremely active area of research.

These few examples reveal the practical applications of some of the emerging scientific views regarding the actions of chemicals that induce cancers in animals. The development of information to test the LNT hypothesis (the standard default) is an important area of research that can not only improve the basis for risk assessment, but can also more generally advance knowledge of carcinogenic processes.

## A miscellany of new challenges and approaches

### Uncertainty factors

As introduced in Chapter 8, uncertainty factors (UF) are commonly invoked by risk assessors to deal with uncertain knowledge regarding, for example, differences in response between animals and humans (interspecies extrapolation), and variability in response among humans. Typical defaults for these two sources of uncertainty are UFs of 10.

Table 9.1 *Subdivision of 10-fold uncertainty factors into pharmacokinetic (PK) and pharmacodynamic (PD) contributions*

See text for discussion of how these PK and PD factors can be used. The Sources and recommended reading section provides citation to the basis of these figures in the work of A. Renwick.

| Interspecies differences | | Variability among humans | |
|---|---|---|---|
| PK | PD | PK | PD |
| 4.0 | 2.5 | 3.2 | 3.2 |

If we attribute these inter- and intraspecies differences to a combination of pharmacokinetic and pharmacodynamic differences (and there are no other imaginable reasons for them), then it is of interest to understand the contributions of each source. A substantial effort towards acquiring understanding in this area was spearheaded in the 1980s by Andrew Renwick of the University of Surrey, UK. Renwick carefully evaluated large bodies of literature on the pharmacokinetics and pharmacodynamics of specific substances, and systematically evaluated the contribution of each to observed differences in toxic response in humans and animals, and among humans. He suggested some generalizations could be drawn, and he expressed those generalizations by showing the contribution of each difference towards the total factors of 10 (Renwick did not investigate the empirical basis for the total factors). His findings supported the contributions shown in Table 9.1. These factors in the table can be seen as a new and more refined set of defaults.

The Renwick work can be applied in the following way. Suppose it were possible in a specific case to develop a reasonably thorough picture of the comparative pharmacodynamic characteristics of a compound in humans and rats, and that the work revealed that no difference in pharmacodynamic response (at comparable doses) was expected. We would then turn to Table 9.1 and see that the typical pharmacodynamic difference between humans and animals (the default) puts humans at 2.5 times greater risk than animals. But now in our new case, the difference is seen to be a factor of 1.0 (no difference). We should be allowed to reduce the overall UF of 10 to a factor of 4.0, which is the default for pharmacokinetic differences (which we have not studied). Data substitute for defaults. Use of the Renwick defaults allows us to make some headway without having to take on the heavy burden of investigating both pharmacokinetic and pharma-codynamic efforts. Also, in the example given, the data were sufficient to reduce the default from 2.5 to 1.0; in other cases it would be quite possible to find that the 2.5 UF is too small. Data, whatever they show, should substitute for defaults, assuming they are truly adequate to answer questions about inter-and intraspecies differences.

Regulators in the United States and in Europe, and other public health institutes, appear to have found UF refinements such as those proposed by Renwick to be valuable additions to the risk assessment model for threshold agents, and we should not be surprised to see many examples in the future in which data are judged sufficient to replace the default UFs in just the way we have described.

### Unusual dose–response relationships

More complex dose–response relationships than we have encountered thus far, while not new, have come into increased prominence in the past decade. One type pertains to substances we recognize as *essential nutrients*, and its importance is not in dispute. The second type is said to describe a highly interesting phenomenon called *hormesis*, but its importance is less clear.

We cannot live without the essential nutrients. The absence of vita-mins and minerals from our diet threatens survival. Thus, the health risk of nutritional inadequacy (which takes many forms, depending upon the nutrient) is large at zero intake, and then decreases as intake increases; risks are minimal when nutritionally adequate intakes are achieved (there are several measures of adequacy, not described here).

As intake increases above the range of adequacy a region will be reached at which the adverse effects of excessive intake will begin to manifest themselves. Figure 9.1 depicts these interesting dose–response curves, and the curve at the right side of the figure represents a typical dose–response relationship for toxicity, in this case caused by excessive intakes of substances we cannot live without at lower doses.

The figure comes from a series of reports issued by the Institute of Medicine over the past decade. The experts who authored these reports revisited the question of recommended daily allowances and other measures of nutrient adequacy, and made recommendations regard-ing macronutrients (proteins, carbohydrates, fats and oils) and for micronutrients (vitamins and minerals). The Institute has had a long history of developing recommended intake levels, but in the recent

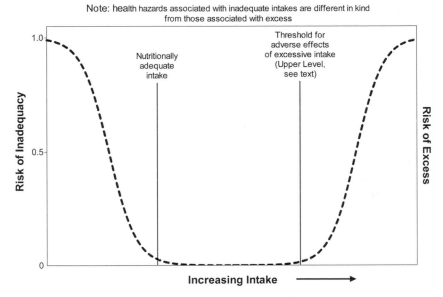

Figure 9.1  Dose–response relations for essential nutrients.

string of reports additional analysis was provided concerning excessive intakes. This new analysis, which models that used for threshold toxicants, was developed because of the trend toward increasing nutrient supplementation of the diet, and the possible consequences of "too much of a good thing." The Institute recommendations for so-called upper levels (ULs, Figure 9.1) for essential nutrients have not yet been translated into government policy regarding food and supplement labeling, but the effort is underway around the world.

Hormesis is a subject of potentially broader scope than the essential nutrient matter, but has not yet had a role in the formulation of public health and regulatory policies concerning hazardous substances. In the crudest of terms, hormesis is said to describe a phenomenon in which toxicants display their usual dose–response characteristics over a range of high doses (from the NOAEL upwards), but then are seen to offer what appears to be a health protective effect over a range of experimental doses below the NOAEL. Thus, rather than showing no response different from the control response in the range from zero dose to the NOAEL for toxicity, there is an observable response in which the test compound shows less toxicity than is associated with the level of toxicity (the "background" of toxicity) displayed by the control animals: high-dose toxicity, low-dose health protection (both measured relative to controls). A little of a bad thing is good.

It seems that large numbers of chemicals, in equally large numbers of test systems, from mammals to insects, vertebrates to invertebrates, microorganisms to plants, exhibit hormetic dose–response relationships. The relationship is not the same as that described earlier for nutrients, in two ways. First, in the case of hormesis the biological response – the toxicity endpoint – is the same in the protective region and in the region of toxicity (i.e., liver cancer incidence is reduced relative to control incidence over a range of low doses, and then as the NOAEL is exceeded, liver cancer incidence increases above that of controls). This is true hormesis.

In the case of a nutrient there is a low-dose adverse effect due to nutritional inadequacy, but the nature of the adverse effect is completely different from that which becomes manifest as the region of high-dose toxicity is entered. Also, the very large risk associated with severe nutrient deficiency at doses near zero is not at all present in the case of hormesis.

The phenomenon of hormesis has been investigated and reported most extensively by Edward Calabrese and colleagues at the University of Massachusetts in Amherst. Calabrese began in the early 1990s investigating, through careful review of thousands of published dose–response curves, what is in fact a rather old idea – that a small amount of stress to biological systems results in healthful responses (think of homeopathic medicine) – and has documented large numbers of cases. There seems to be widespread acknowledgment that hormesis is a real phenomenon, and equally widespread acknowledgement of its unexplained nature. If hormetic dose–response relationships were accepted as a general phenomenon, we would have to rethink some of the fundamental precepts guiding our current risk assessment approaches. The models for dealing with low dose risks currently do not acknowledge the possibility that we do not have to drop very far below the NOAEL before we encounter doses that are not only not risky, but positively beneficial!

There is a small but growing number of scientists engaged in new experimental research on this topic, and publications describing the phenomenon seem to be appearing with increasing frequency. It does not seem to have gathered sufficient momentum to engage the strong interest of regulators, and there is no indication that the latter are ready to reject their current approaches to risk assessment and adopt a hormesis-based model. Part of what hinders such acceptance is the almost certain view among public health and regulatory officials that experimental data are useful and important to establish toxicity, but nothing short of human clinical trials could possibly document a reduced risk of toxicity (a health benefit). No human data of this nature have been developed. Calabrese acknowledges this difficult impediment, but is quick to note that low dose health benefits do not have to be incorporated into a hormesis-based risk model. Such a model would only need to incorporate the more certain notion that risk falls off far more quickly than is allowed for by our current risk models.

Hormesis will be a subject of steady and perhaps increasing interest, but whether and in what way it moves to the center of the policy stage for toxic substances is beyond anyone's predictive powers.

### Endocrine-disrupting chemicals

That some pharmaceuticals and chemical products can somehow mimic the actions of the body's various hormones is well established, as is the fact that such actions can have significant adverse health consequences. Much effort is now devoted to studying the effects of such chemicals, although the appropriate methods for their study are still under debate. Many toxicologists studying endocrine system effects of chemicals focus on what some have termed "the fragile fetus" and the neonatal stage of life, and some contend that exposure to endocrine-disrupting substances during these highly sensitive periods of differentiation may have multiple health consequences that do not become fully realized until later in life. The once-used pharmaceutical agent DES (Chapter 6) has a chemical structure that can mimic natural estrogen, and the effects of such activity can be highly detrimental. They include vaginal cancer in young women exposed in utero and other delayed manifestations of its powerful estrogenic effects. No similarly potent endocrine-disrupting chemical has been identified, but concerns have arisen over some commercial products.

Diethylstilbestrol

The term "phthalates" refers to a group of esters of phthalic acid, some of which are widely used to "soften" plastics such as polyvinylchloride and to serve several other commercial purposes. DEHP, mentioned earlier, is one of several compounds in this class. These so-called "plasticizers" can be present at relatively high levels in some plastics, and can migrate in small amounts out of the plastic (for example, into children who may chew on soft plastic toys or into foods and beverages present in certain plastic containers). The rodent carcinogenicity of some phthalates has been known for some time, and much evidence supports the role of peroxisome proliferation in the production of those tumors (see above). Developmental toxicity is another characteristic of some phthalates. Human exposure is widespread; recent NHANES surveys, described earlier, have turned up evidence of phthalate exposures in 75% of urine samples taken. Interest in phthalate toxicity has recently been intensified by the publication of two investigations in humans, one suggesting adverse effects on sperm motility, the second suggesting an association between prenatal phthalate exposure and reduced anogenital distance in male infants (an indication of an anti-androgenic effect also seen in rodent studies). While neither of these findings is at all definitive, they do heighten concerns.

Di(2-ethylhexyl)phthalate

Bisphenol A

Bisphenol A is present in certain plastics used as food and beverage containers. The EPA has established an RfD for this important product at 50 μg/kg/day, based on data from toxicity studies published in the 1980s.

Current human exposures seem to be well below the RfD. But in the time since the RfD was published many investigators have pursued research into the compound's reproductive and developmental effects, and some report biological activity said to be related to endocrine system disruption at doses below the RfD (which was derived by the application of large uncertainty factors to the earlier toxicity studies). Either the newer test systems are yielding irrelevant toxicological findings, or they are telling us that more traditional study protocols fail to uncover many important endocrine system effects. Many investigators are in hot pursuit of answers.

Much of the work on the so-called endocrine-disrupting chemicals involves the use of research models that have not undergone the type of validation that has established the reliability of the types of testing protocols we described earlier in Chapter 3. This is not in itself a reason to ignore these findings, but it no doubt explains why regulators do not seem to be rushing to use such data for risk assessments. At the same time, agencies around the world are searching for the best methods to study the toxicity of this class of agents. The topic remains unsettled and controversial; whether these various signals from some segments of the toxicology and epidemiology community are of high public health importance, or are mostly false alarms, will not be clear for some time to come.

## Toxicology at the nanoscale

The Nobel laureate Richard P. Feynman described, in a lecture delivered in 1959, the future of miniaturization. The published version of his lecture is called *There's Plenty of Room at the Bottom* and in it can be found a recipe for putting the entire Encyclopedia Britannica on the very small head of a very small pin. Feynman's comments set into motion an entirely new area of study and have lead to what have become known as the fields of *nanoscience* and *nanotechnology*. Chemists, physicists, materials scientists, and engineers have come together over the past several decades to produce with high accuracy and precision materials that have dimensions measured in nanometers (nm, $10^{-9}$ meters, about 1/100 000 the width of a human hair). Specifically, materials with one, two, or three dimensions of 100 nm or less (called, respectively, nanofilms, nanotubes, and nanoparticles) qualify as products of nanotechnology. It appears that almost any chemical substance that is a solid under ordinary conditions of temperature and pressure can be manufactured ("engineered") to dimensions in the nano range. Special techniques are necessary to accomplish this engineering; indeed, when these materials are produced under ordinary conditions, as so-called "bulk" products, most of their particles have dimensions in the "macro" range, perhaps 10 000 nm or greater. What has been clearly established is that a material produced with nanodimensions takes on properties that are different, sometimes dramatically so, from the chemically identical material produced as a bulk material. Many of these "new" properties – electronic, magnetic, catalytic, and mechanical – create opportunities for improving everything from drug delivery systems to consumer products of many types, and materials used in many industries. Some nanomaterials also seem potentially to be useful in remediation of hazardous waste sites (by providing for highly efficient catalytic degradation of contaminants), and even in improving the efficiency of chemical manufacturing (by reducing wastes, among other things). The list of current and potential applications is very long, and patents are issued daily, not only in developed but also in many developing countries. Some have estimated that by the year 2014, 15% of global sales revenues will be due to products of nanotechnology. A discussion of just how nanomaterials provide their many benefits is well beyond our scope, but we can say that some of the beneficial properties of substances produced at the nanoscale may also signal the existence of a downside – as yet hidden but with potentially significant human and environmental health risks.

One hint of possible trouble to come is provided by the information we described in Chapter 4, related to airborne particulate matter (PM). The available evidence ascribes significant increases in the risks of asthma and other respiratory diseases, certain cardiovascular conditions, and lung cancer to PM exposure, particularly those that average less than 2.5 μm (2500 nm) in size. As we noted, the chemical composition of these particles varies widely, depending upon source, but may not be as important as particle size as a risk determinant. Moreover, there is some experimental evidence pointing to the so-called "ultrafines," PM with dimensions below 100 nm, as significant contributors to PM risk. In addition some experimental studies have demonstrated that ultrafines not only distribute themselves throughout the airways, but seem to be able to "translocate" to other parts of the body – liver, heart, perhaps the CNS.

Proponents of nanotechnology argue that the ultrafines present in PM are not likely to represent engineered nanoparticles. The former are chemically diverse and have a wider range of size and shape, whereas the latter are uniform chemically, and in terms of size and shape. This argument may be correct, but so little experimental study on the toxicity of engineered nanomaterials has been developed that it is not yet possible to confirm this.

Some of the physical attributes of nanoparticles attract the interest of toxicologists. The surfaces of these materials are truly enormous relative to their mass, and one consequence of these surface properties is extreme catalytic power. Toxicologists can easily envision that such a property can create significant cellular damage. Indeed, some particles with dimensions of just a few nanometers can readily escape the immune system's detection power and survive to reach cells, pass through cell membranes (10 nm dimension), and persist intracellularly, perhaps causing chemical damage the whole time. Imagine a particle with a super-reactive surface reaching the synaptic junction (they are 3–40 nm spaces) and disrupting nerve transmission. None of these possibilities has been studied, but they may signal concerns for unusual (perhaps unprecedented?) toxic responses.

The pace of nanomaterials production and research into new applications is greatly in excess of the pace of toxicology testing and risk assessment. Although there are compelling reasons to believe that human exposures, whether occupational or consumer, are likely to be small (nanoparticles tend to agglomerate into larger particles when released into the air), actual evidence on this question has not been developed.

And there are significant questions (as yet unanswered) regarding methods for assessing risks. Perhaps, for example, traditional notions of dose–response relationships are inapplicable when the particle size, or perhaps the surface area (huge relative to mass) is the real risk determinant. There is much to be done, and those promoting these exciting new products should no doubt be equally determined to promote the development of the information needed for reliably assessing their health and environmental risks.

## Microbial pathogens

While we remain anxious about the health threats associated with the world of industrial chemicals, we perhaps should be even more anxious about those agents of disease that have been with us since we first risked walking the earth. Infectious diseases, those passed to us from other people or from animals, are major health burdens, and on a worldwide basis are far more serious than those we have been discussing in this book. This conclusion may be wrong, because it may be in part based on the ease with which we can develop statistics on disease burdens; it is far easier to measure the disease burden related to infections than it is to measure the total burden that might be related to chemicals. In any case, it seems relatively clear that infectious diseases and chronic diseases related to poor nutrition and to smoking account for the greatest share of disease-related morbidity and mortality. One component of this share can be related to food- and water-borne *microbial pathogens*. The latter are typically bacteria that, under certain conditions, can produce toxins and thereby cause illness. The problematic species are not infectious agents in the usual sense, that is, they are not transmitted from person-to-person. Rather, they can, under some circumstances, contaminate food or water. In some cases, the pathogens grow and produce toxins in foods; in other cases, the pathogen is swallowed, comes to infect the human gastrointestinal tract, and while growing there produces its toxins. In either case, the result can be what we call "food poisoning." "Water poisoning," while a major public health problem in many areas of the world, can easily be avoided through chlorination – the simple, highly effective technique discovered in 1919 by Abel Wolman, then of the Maryland Department of Public Health, and for many years a professor at the Johns Hopkins University.

Food-borne pathogens are not so readily treatable, and disease outbreaks (a term used to describe food poisoning events that have a single source and that affect many persons at one time) are not uncommon. Individual cases also occur, but these are not so easily countable. In any case, the total annual burden of disease from food-borne pathogens in the United States is said by the CDC to be in the tens of millions. Most of these cases amount to nothing but the extreme discomfort associated with vomiting and diarrhea, but mortality can be high in certain subpopulations – children, the elderly, those with impaired immune functions.

Bacterial pathogens can enter the food chain very early; in fact many foods leaving the farm carry them. But they become dangerous only if foods are mishandled, and conditions for their growth and for toxin production are created. They can also be destroyed by heat and other food processing techniques. And, of course, they may be destroyed by cooking. So in the whole chain from "farm-to-table" there are opportunities for both the growth and the destruction of food-borne pathogens. The major culprits are certain species and subtypes of *Salmonella* and *Shigella*, *Listeria monocytogenes*, *Escherichia coli* 0157:H7 (all causing infections), and *Staphylococcus aureus* and *Clostridium botulinum* (both capable of toxin production in foods). The toxins produced in food are proteins of extreme toxicity (recall Chapter 4), but they are also vulnerable to ready destruction by heat.

Successful development of microbial risk assessment models is a leading priority in public health and regulatory agencies. In the United States the effort is led by the Food Safety and Inspection Service of the Department of Agriculture and the FDA. These agencies oversee food safety problems related both to chemical and microbial agents. Microbial risk assessment activities in the EU and at the UN's Food and Agriculture Organization (FAO) and the WHO are also now quite visible. The ultimate forms these models will take are unknown, and it is too soon to say just how the results of microbial risk assessments will affect regulatory and public health policies. We can highlight some of the challenging issues that are being scrutinized.

The risk assessment framework we have described for chemical toxicity is applicable to microbial risk assessment. Once the information is available on microbial hazards, which are for the most part acute (immediately observable) conditions resulting from acute (one-time) exposures, and their dose (pathogen count)–response characteristics, we should be ready to assess the risks associated with any dose of interest. Hazard information for the important pathogens is readily available but, as expected, their dose–response characteristics are much harder to come by. So with pathogen risk assessment we see the same types of uncertainties creeping into the framework as we have encountered for chemicals.

But these sources of health harm have a characteristic that is not associated with chemicals – they are living organisms that can reproduce. So, in their journey from farm to table, organism counts can increase and, if conditions for growth are unfavorable, they can decrease. Some chemicals may undergo a degree of destruction as they move from source to target, but this phenomenon is far more problematic and unpredictable in the case of microbes.

Microbiologists have developed ways to model microbial growth and, using assumptions related to the expected behavior of organisms under different environmental conditions, these models are then coupled with dose–response models with the result that risks (responses) can be estimated, given a certain degree of knowledge about initial microbe counts and the environmental conditions (related to food processing) these microbes encounter as they move from these initial conditions to the dinner plate.

Interestingly, most microbiologists seem to hold to the "no-threshold" hypothesis for those pathogens that can infect. A single bacterium, if still viable when it reaches the gastrointestinal tract, can multiply and cause disease. Sigmoid (S)-shaped dose–response models, with no threshold characteristics, are typically proposed for microbial risk assessments (though threshold models are used for those toxins produced in foods). The discipline of microbial risk assessment is in development, and practitioners have not reached consensus on all features of the process. Defaults are used, but there appears to have been little of the type of science-policy discourse on this topic that has informed chemical risk assessment. This will no doubt come as the discipline matures, and, in the meantime, it is exciting to observe the worldwide scientific dialogue that is underway.

We have offered just a hint of some of the new challenges and approaches to them that now occupy the agendas of public health and regulatory authorities, as well as those of the industries they regulate and the public they serve. We have omitted far more than we have included but it is clear that risk assessment and the various research efforts that support it continue to be seen as central to public health improvement. Of course that improvement can be expected only if the health risks we uncover are subjected to appropriate and effective risk management policies. That is the subject to which we turn, in Chapter 11, after a brief chapter on risks that are "managed" not by regulators, but by judges and juries.

# 10

# Risk assessment IV: the courtroom

In the 1970s, not long after the discovery that the synthetic estrogen DES could cause a rare form of vaginal cancer in young women exposed in utero, lawsuits on behalf of the victims began to be filed in federal and state courts, seeking payment for medical damages from the manufacturers of the drug. In the same time period similar lawsuits began to be filed on behalf of individuals who believed they had been harmed by exposure to asbestos. Veterans of the Vietnam War began, in the early 1980s, bringing similar suits against manufacturers of a herbicide mixture called Agent Orange that had been used widely and intensively by the US military to clear forests in Vietnam. Many veterans claimed that the massive spray applications of the herbicide, which contained low levels of dioxins, caused them to experience harmful exposures, and that they and their offspring suffered a range of serious diseases as a result. Although such lawsuits existed prior to the 1970s and 1980s, the three involving DES, asbestos, and Agent Orange initiated the age of the "mass tort," and it continues to this day.

## Legal rights

Under our laws individuals have the right to bring lawsuits against parties whose actions they believe have caused them harm. Harm (a "tort") can come in many forms, but here we are concerned with harm that may have been caused by a chemical (a drug or a consumer product, for example) or by exposure to a chemical contaminant in the environment. The former are usually referred to as product liability cases and the latter have come to be known as "toxic tort" cases. Most of the asbestos-related litigation arose because of occupational exposures, and many other tort cases have stemmed from exposures incurred on the job, but litigation has also arisen over exposures arising from the general environment.

While individuals who have diseases or other types of medical injuries have a right to sue manufacturers, they also have the burden of providing evidence that they were in fact exposed to a harmful substance, and that they were exposed at a sufficiently high level, for a sufficient period of time, to make it "more likely to be true than not true" that the injury or disease they have was caused by the substance. They also have the burden of identifying the specific manufacturer(s) or distributor(s) or user(s) of the substance that caused their injury or disease. These are heavy burdens indeed, and we shall be discussing the types of evidence that are typically necessary to meet them.

If those bringing lawsuits – the plaintiffs – can make a convincing case for causation, then juries may require the parties that caused the injury – the defendants – to compensate them for the harm they have incurred. Moreover, if plaintiffs can demonstrate convincingly that the defendants' actions were in some way negligent, then additional compensation in the form of "punitive damages" may be required. Negligence may, for example, relate to inappropriate release of a chemical to the environment, failure to warn workers or consumers about the possible dangers of a substance, or even failure to test the product adequately before its introduction to the market. Generally, regulatory approval of a substance does not ensure a manufacturer's protection against lawsuits or negative jury decisions.

Complex legal issues of many types attend these types of cases; we need not deal with them here (also, I am not competent to deal with them). What we can deal with are the scientific issues, in particular those relating to the question of causation. Both plaintiffs and defendants will engage experts to evaluate the facts regarding exposure, and develop opinions regarding the probability that a plaintiff's injury or disease was caused by the incurred exposure: for the plaintiff to be successful, the plaintiff's experts must demonstrate that the available evidence supports a "more probable than not" criterion; a successful defense requires experts to demonstrate a contrary view. If we ignore that category of "expert" who is always ready to fashion novel hypotheses that allow him or her to argue any position that will help his client, be it plaintiff or defendant, we can focus on the types of evidence and the methods used to evaluate the evidence, that are useful and relevant to pinning down disease causation in individuals.

The types of evidence and methods used in these evaluations have some similarity to those used to conduct risk assessments for regulatory purposes. Indeed, the general risk assessment framework used by regulators is, we shall show, an appropriate one for organizing evidence and for reaching decisions regarding the likelihood that an individual was harmed by a given exposure circumstance. But, as we shall see, because we are dealing with a specific disease or injury in a specific individual, the types of evidence we turn to may be somewhat different from that used by regulators. Moreover, regulatory defaults are not particularly relevant; such defaults allow regulatory risk assessments to proceed in the absence of knowledge, to ensure that actions to protect public health are not stymied. Although the legal right to sue is in part based upon the societal goal of discouraging inadequate or negligent behavior on the part of corporations, the scientific judgment regarding causation is not expected to be influenced by the kind of "science policy" criteria (which include a significant element of caution) used to develop regulatory defaults. As we outline the risk assessment approach used in evaluating personal injury claims we shall highlight a few of the principal differences from that undertaken in the regulatory context.

At the outset, it should be mentioned that the scientific and medical professions have not devoted significant, collective energies to the problem of personal injury assessment, and there appears to be no single methodology that can be said to represent a "standard" in the field. We present here what may be a majority view, but to call it a consensus view would be misleading. It is nevertheless worth discussion because toxic tort and product liability cases are highly important and growing elements of America's judicial scene, and more and more scientists and physicians are being asked to become involved as experts.

## Causation in individuals: the problem

Mr. Z is a 50-year-old white male who has just been diagnosed with leukemia. Mr. Z also lives in a community in which several industrial chemicals have migrated from a nearby industrial site, and are present in the community's ground water. Mr. Z has a well in his yard and has been consuming the well water from the contaminated aquifer for about 20 years. When the contaminants first entered the ground water system is not known for sure, but experts believe they have been present for 10–15 years. Data exist on the current levels in individual wells. When the contamination was first discovered, about one year prior to Mr. Z's diagnosis, individual wells were closed and the community was provided with uncontaminated municipal water supplies. Mr. Z believes his leukemia was caused by well water contaminants, and brings a lawsuit against the corporation that owns and operates the nearby manufacturing site. In this he is joined by 100 neighbors who believe many of the maladies they suffer are due to the failure of the operators of the manufacturing site to control the environmental release of its wastes.

This is the kind of situation that an environmental regulatory agency might investigate, although the agency would not direct its attention to the question of Mr. Z and his leukemia, or to the similar questions raised by the other plaintiffs. The regulatory agency's objective is to understand whether the community as a whole is at excess risk because of the ground water contamination. A step toward that objective is to conduct a typical regulatory risk assessment. As we have noted in Chapters 7 and 8, the goal of the regulators is to consider the chemicals present, their toxic hazards, and dose–response characteristics, and also to evaluate exposures, usually focusing on the "high end." Defaults are employed, both with regard to toxicity and dose–response assessments and with respect to exposure assessment. The regulatory question is typically directed at acquiring an understanding of the risks the population might face if they were to continue drinking the contaminated water. If total risks exceeded some risk management targets, the agency might mandate remediation of the ground water, to achieve a suitably safe supply for the future. The toxicity endpoints evaluated may or may not relate to those complained of by the individual plaintiffs; regulators use animal toxicity data or, when it is available, epidemiology data, and focus (as we have noted in earlier chapters) on the particular endpoints that are the most sensitive indicators of adverse health effects. Generic uncertainty factors and cancer slope factors are used; none of these will be known to relate in any way to specific people. Regulatory risk assessments focus on generic (hypothetical) individuals, who are equally sensitive to the effects of a chemical, and who are identical with respect to all of the factors that affect their exposure. The generic population is also likely to represent the "high risk" end of the distribution of risks in the population (most sensitive to toxicity, and also incurring the high end of the exposure range). There is no basis for believing that the risks described with the regulatory model can be shown to be applicable to any specific people in the community.

An agency such as the ATSDR or a state public health agency might be called in to conduct a health survey, or even an epidemiology study, in the community. Such a study may be informative at the population level (e.g., it may reveal that there are greater numbers of leukemia cases than would be expected), but there is likely to be insufficient information on disease rates in the community and in comparison communities to allow evaluation of many medical conditions. Even if some may be found at higher than expected rates, there may not be a clear-cut relationship to the magnitude or duration of contamination exposure. It is, as we have stated several times, nearly impossible to establish causation with a single investigation, and this problem is significantly exacerbated when the investigation concerns an environmental (as opposed to an occupational) setting. Even if the excess numbers of leukemia cases were considered to be strongly associated with, or even causally linked to, the contamination, this does not mean that each of the individual leukemia cases, including Mr. Z's, was, in fact, caused by the contamination. So what is Mr. Z to do?

## General and specific causation

Let us assume that Mr. Z does indeed have leukemia. For many conditions claimed by plaintiffs, especially those that are highly subjective in nature (headaches, nausea, intermittent skin rashes, insomnia, muscle pain), a similarly objective diagnosis may not be possible; this creates many problems in causation evaluation which we shall not try to cope with here. But to evaluate the likelihood that Mr. Z's leukemia was caused by one or more water contaminants, it will be necessary to determine whether there is evidence in the scientific literature that is sufficient to establish a causal link (in the sense, for example, described by IARC and discussed in Chapter 6) between exposure to any one of those contaminants and leukemia. This evaluation is referred to as an analysis of *general causation*. Thus, it is directed at the question of whether one or more of the chemicals to which Mr. Z was exposed is known, in a general sense, to be a cause of leukemia. If benzene is, for example, one of the chemicals found in Mr. Z's well, and it can be established that he consumed water containing benzene, then we could conclude that general causation is established.

But in establishing general causation we have taken only the first step. To establish that Mr. Z's specific disease was caused by the benzene in his well water, then it would be necessary to show that he incurred sufficient exposure, measured both as magnitude and duration, to make it more likely that his leukemia resulted from the benzene exposure than from whatever other factors in his life might have been responsible. While benzene is a cause of leukemia, it is neither a necessary nor a sufficient cause. Mr. Z reached age 50 and at that age had a risk of developing leukemia even if he had never experienced benzene exposure (at least at levels above that low level all of us experience from benzene's ubiquitous presence in ambient air). If Mr. Z wishes to blame the corporation responsible for his excessive benzene exposure, his expert epidemiologist or toxicologist will have to make a judgment regarding the adequacy of that excess exposure to increase Mr. Z's risk to a level that is sufficient to support a finding that it is more likely the disease derived from the benzene exposure than it did from whatever factors contributed to his normal risk. Here, on the question of *specific causation* (causation specific to Mr. Z), scientific squabbling usually begins (actually, it usually occurs at all levels of evaluation, but it can be particularly troublesome on this point).

Let us assume that enough information is available regarding the levels of benzene in Mr. Z's well, the number of years he consumed the water, and even his water consumption rate, to derive a reasonably accurate estimate of his cumulative exposure from this source. The epidemiologists and biostatisticians carefully evaluate the dose–response data from the published epidemiology studies used as the basis for classifying benzene as a cause of leukemia. Further assume that we learn from this evaluation that Mr. Z incurred a cumulative benzene exposure approximately equivalent to the cumulative exposure that was found to cause a three-fold excess risk of leukemia in the occupational studies of benzene exposure. A relative risk of three.

We might then assign Mr. Z's normal background risk of developing leukemia a value of one (this is not a real risk number, but we are merely using it as a point of comparison). Now we have learned that Mr. Z incurred sufficient benzene exposure from his well to triple that background risk; his total risk is thus four, and three units of that came from the well exposure. We would thus have to say there was a 75% chance that his leukemia came from the well water, and was not due to whatever factors there are that contribute to the normal background risk. A 75% probability exceeds that legal criterion of "more likely than not," which suggests that any probability greater than 50% (sufficient exposure to at least double the background risk) would be adequate and necessary to demonstrate causation, as measured against the typical legal criterion.

Of course if Mr. Z's cumulative exposure were found to be insufficient to at least double his normal background risk, he would fail to make his case that the corporation was to blame.

Even when Mr. Z's expert makes the case for specific causation, it remains possible that there are other medical and exposure issues that make alternative causes even more likely than benzene. Possible alternative causes are always in the spotlight, at least among defendants.

Although this sketch might sound like a reasonably objective approach, it is by no means the only method of evaluation that makes its way into the courtroom.

## Controversies

In its broad outline the evaluation described follows the same risk assessment framework used by regulators, but gives different weights to different types of scientific evidence. As described, it requires that general causation be based upon solid epidemiological evidence. Animal data may be useful to buttress epidemiological evidence, but standing alone, in the absence of adequate epidemiological data, animal data would be considered by many toxicologists as insufficient to establish general causation. Regulators use animal data all the time in risk assessment, without regard to the question of whether such data truly establish causation; they operate under the assumption that such data are sufficiently predictive of some type of adverse effect in humans, but are not concerned to establish, before acting, just what that effect is with the degree of certainty necessary, in a tort setting, to establish general causation.

But many toxicologists will argue that animal data are highly predictive, and offer arguments that what we are evaluating in the tort context is really not very different from what occurs in the regulatory risk assessment process. Arguments for and against causation come in numerous forms, and judges and juries are truly challenged when attempting to sort good and relevant evidence and scientific methods from inadequate evidence inadequately evaluated. It is a problem that is not likely to disappear without greater attention from the scientific and medical communities.

Moreover, the "risk-doubling" criterion given in the example of Mr. Z is by no means a universally accepted standard, and quantitative evaluations of this sort are often avoided altogether by some experts.

# The Daubert decision

In 1993 the United States Supreme Court ruled on a case involving a drug called Benedectin, used to alleviate morning sickness in pregnant women. The drug came under attack during the 1980s because of allegations that it increased the risk of birth defects. Lawsuits ensued. Various challenges to the evidence proffered by plaintiff experts were mounted by defendants, and one of the cases (Daubert) made its way to the Supreme Court. In the end the court agreed with defendants on the specific case, but its opinion included a number of "general observations" regarding the nature of reliable scientific evidence and the admissibility of scientific evidence in courtroom proceedings.

In a much earlier judicial decision related to the scientific reliability of lie detection tests, the so-called "Frye rule" had emerged. The rule emphasized the need for scientific evidence to have "general acceptance" before it could be presented to a jury; it held sway in courts until the Daubert decision (and is still considered appropriate in some jurisdictions).

The Daubert decision led to a new set of criteria to be used to judge the reliability of evidence presented in these cases. Most importantly, the Daubert decision contained a provision that put judges into the role of scientific "gatekeepers." Judges, the court said, should review evidence to be offered by experts, and make a decision, based on the court's criteria for evaluating reliable scientific evidence, on whether those experts should be allowed to testify before a jury. Under the Daubert rule, the judge has become a kind of scientific peer reviewer. So-called Daubert hearings, in which scientific experts can be challenged with respect to credentials, the nature of the evidence they would propose to present, and the methods used for evaluation of that evidence, have become common features of the judicial scene.

Daubert is controversial. Many judges are understandably hesitant to play the peer review role. In fact the Supreme Court decision contained a minority view that strongly argued against putting judges in such a position. Many ask why experts with bona fide credentials should not be permitted to offer views that may not strictly fall within the mainstream; although others argue that some mechanism needs to be in place to eliminate truly unsupported opinions, based on inadequate methods of evaluation, that may nevertheless sound plausible to juries.

The relatively small amount of space given to this subject is significantly out of proportion with its social and scientific importance. But it may accurately reflect the amount of attention given to the subject by scientific professionals. Most of the "scientific" literature on the topic appears in legal journals. It is a subject that deserves far more scientific scrutiny than it has yet received.

# 11

# The management of risk

Federal and state legislators in the United States have enacted laws that mandate certain types of controls on human exposure to just about every category of consumer and medical product, industrial chemical, and environmental pollutant. Although these statutes vary regarding the extent of required risk control and the factors that need to be considered in risk management decisions, all seek protection of human health. Table 11.1 contains a list of some of the major laws, the categories of chemicals they cover, and a further notation regarding the factors that managers need to consider when making decisions. Countries in much of the rest of the world have enacted or are enacting similar legislation.

The first thing worth noting about these laws is that they do not treat all sources of chemical exposure in the same way. Congress has directed the FDA, for instance, to consider only whether an added food ingredient is "safe" when making decisions about allowing it to be used, whereas the same agency is allowed to balance the risks associated with a new drug against the health risks that might exist if the drug were not available for use in disease treatment; this balancing is often called a risk–benefit analysis. Costs are not allowed to be considered by the agency either in the case of drugs or of food ingredients. Under some laws the EPA is allowed to consider both health risks and the availability and costs of technology to control these risks, whereas others require the agency to consider only health risks when making decisions. The OSHA, in setting workplace limits, is supposed to ensure worker safety and also consider the availability of technology to control exposure. The notations in Table 11.1 – risk only, technical

Table 11.1 *Some federal laws under which chemicals are regulated in the United States*

| Law | Regulatory agency | Regulated products | Regulatory model[a] |
|---|---|---|---|
| Food, Drug and Cosmetics Act | FDA | Foods, drugs, and cosmetics, medical devices, veterinary drugs | Risk (food, cosmetics) Balancing (drugs, medical devices) |
| Meat and Poultry Inspection Act | USDA/FSIS | Meat and poultry | Risk |
| Federal Insecticide, Fungicide, and Rodenticide Act | EPA | Pesticides, non-food uses | Balancing |
| Food Quality Protection Act | EPA | Pesticide residues in food | Risk |
| Federal Hazardous Substances Act | CPSC | Household products | Risk |
| Occupational Safety and Health Act | OSHA | Workplace chemicals | Technical feasibility |
| Clean Air Act | EPA | Air pollutants | Technical feasibility |
| Clean Water Act | EPA | Water pollutants | Technical feasibility |
| Safe Drinking Water Act | EPA | Drinking water contaminants | Technical feasibility |
| Resource Conservation and Recovery Act | EPA | Hazardous wastes | Technical feasibility |
| Toxic Substances Control Act | EPA | Industrial chemicals not covered elsewhere | Balancing |

[a] "Risk" means the agency considers only risk information when reaching decisions. "Balancing" means that both risks and benefits are considered. "Technical feasibility" means that the law requires the agency not only to consider risks, but also the availability and costs of technology to control risk. Some laws invoke more than one model.

feasibility, balancing – are provided as a (somewhat oversimplified) guide to what decision-makers are required to take into account.

This is confusing. Why don't risk assessors simply decide what level of exposure is safe for each chemical, and risk managers simply put into effect mechanisms to ensure that industry reaches the safe level? Why should different sources of risk be treated differently? Why apply a "no risk" standard to certain substances (e.g., those intentionally introduced into food, such as aspartame) and an apparently more lenient risk–benefit standard to unwanted contaminants of food such as PCBs, methylmercury, and aflatoxins (which the FDA applies under another section of food law)? Why allow technological limitations to influence any decision about health? What is this risk–benefit "balancing" nonsense? Aren't some of these statutes simply sophisticated mechanisms to allow polluters to expose people to risk?

These are pretty good questions, and they are not easy to answer, especially those concerning the different decision-making criteria associated with different laws.

Let us deal with this last issue first, and simply note that the laws listed in Table 11.1 each have their own history and were generally enacted quite independently. Their particular forms were fashioned out of a complex interaction of industry, consumer and environmental activist, and governmental constituencies that each brought its own agenda to the legislative process. It is not the purpose here to try to understand how these differences came about, but rather to explore some of the effects of these differences on the problem of deciding what limits ought to be placed on human exposures to environmental chemicals.

This chapter is by no means a comprehensive portrait of risk management issues. Many exceedingly complex technical and policy matters, to say nothing of the often volatile political factors, influence decision-making in particular cases. Emphasis here is on certain technical issues that arise in the use of the risk information that has been the subject of this book.

## Safety

It should be clear by now that risk assessors do not know how to draw a sharp line between "safe" and "unsafe" exposures to any chemical. The very notion of "safety" is scientifically wrongheaded, if it is to mean the absolute absence of risk. If "safety" is defined in this way, it becomes in most cases impossible to know when it has been achieved, because to do so requires the proof that something – in this case, risk – does not exist. The term remains important – we still ask about a safe food supply, safe water, and safe household products – but over the past few decades some uses of the term have been replaced with terms that more explicitly acknowledge risk.

That no risk exists can be proved under one and only one set of circumstances: when it is certain that exposure does not exist. How can the latter condition be ensured? The only real way is to guarantee that a chemical is not used for any purpose. We can know that exposure to cyclamate, a commonly used non-nutritive sweetener until 35 years ago when it was banned by the FDA, does not exist (save, perhaps, in somebody's laboratory, where a few bottles might be sitting around, and in some other countries, where its use is permitted), because food manufacturers are prohibited from adding it to foods and beverages. Cyclamate thus poses no risk to individuals in the United States – under present conditions we are absolutely protected from any risks this chemical may pose (a debate still continues about whether it is carcinogenic but that is irrelevant to this discussion).

But such "absolutely safe" situations are not of much interest. While the use of some chemicals can be banned, it is not realistic to expect this approach to be applicable to all industrial chemicals, consumer products, or to the polluting by-products of industrial society. If the goal of absolute safety (zero risk) from these products is desired, then such wholesale banning would be necessary. We do not appear ready to turn back the calendar 200 years.

To further this discussion let us divide environmental chemicals into three broad groups. First there is the enormous group of naturally occurring chemicals that reach us primarily through food and products such as cosmetics, but also through other media. Second are industrially produced chemicals that are manufactured for specific purposes. And third are the industrial pollutants – chemical by-products of fuel use, the chemical industry, and most other types of manufacturing.

Of these three groups the second is probably the easiest to control, in a technical sense. If society wishes to guarantee absolute protection from any of the forms of toxicity associated with these substances, then it will be necessary to prohibit their use altogether. If regulators were successful at banning all the products of industry, some of the pollutants arising as by-products of their manufacture (group 3) would also disappear, but the large number of chemicals arising from other sources of pollution would still have to be dealt with. And, because it is not possible to ban food or the natural world, society would continue to live with the large number of chemicals from this source which, we have noted, pose risks of largely unknown magnitude because scientists have not paid much attention to any except those having serious, acute toxicities.

Because wholesale bans of this type will not occur, then another approach to achieving safety, at least for pollutants, might be suggested. Why not seek the goal of "no detectable" chemicals in the media of human exposure? If automobiles emit various nitrogen oxides, simply ensure that emission rates are sufficiently low so that these noxious chemicals cannot be found in air. If PCBs are migrating from a hazardous waste site, impose limits on that migration so that no detectable PCBs are found in the off-site environment. Control aflatoxin contamination of raw food commodities to ensure none can be found in finished foods. Why not apply this approach to all pollutants (it obviously is not applicable to products)?

This approach may sound pretty good, but it does not make much sense. The "I can't find it, so it must be safe" approach to controlling environmental risks is flawed because it depends upon the operation of a relationship between technical capabilities to detect the presence of a chemical and the magnitude of the health risks it poses. There is no such relationship. Further elaboration of this issue is in order, and a specific example will be useful, because it leads us into the heart of risk-based decision-making.

Diethylstilbestrol, mentioned in Chapters 5 and 9 as a synthetic estrogen that is also a human carcinogen, was used in the United States from the 1950s until 1979 as a growth promoter in sheep and cattle. Small amounts of this drug, added to animal feed or implanted in the flesh of animals' ears, increase feed efficiency, and it was very widely used for this purpose.

Under the law the FDA is charged with enforcing, carcinogenic substances such as DES can be used in food animal production, as long as "no residue" of the drug is found in edible products, in this case beef.

This "no residue" requirement of the federal food law seems to ensure safety. If there is no residue, then there is no exposure, and, it follows, no risk to anybody. Sounds perfect.

But what, exactly, does "no residue" mean? If the applicable food law is examined more closely, we find it actually says "no residue by a method of analysis" approved by the regulatory agency. This linkage of the phrase "no residue" to a "method of analysis" is important and suggests our legislators actually understood the necessity of establishing such a linkage (little chance this was actually the case).

Why such a linkage? As we emphasized in Chapter 2, any analytical chemist will tell you that no method of analysis can ever reveal, under any circumstances, that a chemical is *not* present. If an analysis is performed on beef suspected of containing DES, and the chemical is not detected, the best the analytical chemist can conclude is that the compound was not present above the minimum concentration of the chemical the analytical method employed was capable of detecting. This concentration is called the "limit of detection" and it varies from chemical to chemical, from one environmental medium to another, and from one method of analysis to another.

Until the early 1970s analytical chemists could detect DES residues in beef tissue, specifically liver, at a level of 5–10 parts DES per one billion parts of beef (5–10 ppb). If DES were present above this level it could be detected with existing analytical procedures, but it could not be found if it were present at any concentration from zero to 5 ppb. Under conditions of cattle dosing approved by the FDA, "no residue" of the drug could be found in the late 1960s. The drug could safely and legally be used.

But, as they are always eager to do, research analytical chemists found ways to improve their procedures, and by the early 1970s they could detect DES residues at about 1 ppb and above. Guess what? DES could be found where none was detected with the earlier, less sensitive method, even though the drug was being administered to cattle at the same (approved) dosages.

This result was not surprising. Once a drug, or any chemical, enters an environmental medium, in this case animal tissue we use as food, some amount is going to be present. Although amounts may decline (as the chemical metabolizes or degrades, for example), it is not possible to conclude its concentration ever goes to zero. The best we can do is to search for it with some method of analysis and, if it is not found, conclude that it is "not present above the detection limit" of whatever analytical method we use. If detection limits improve – become lower – it is expected that the chemical will be found where it could not be seen with the earlier, less sensitive detection procedures. The lower the detection limits, the greater will be the frequency of samples found to contain detectable concentrations.

What should the FDA have done with the analytical data on DES? The law permitted use of the drug only if "no residue" could be found.

The agency acted in accordance with the law and initiated proceedings to ban the drug, an action not completed until 1979.

At the same time the FDA was attempting to deal with DES residues, the agency also recognized the fundamental strangeness of the "no residue" requirement. In effect, it said that if a carcinogenic animal drug could not be detected in food, the food was to be considered safe. This is odd, because it defines safety in terms of the capabilities of analytical chemistry. It is not only odd, it makes no sense whatsoever. Our ability to detect chemicals in the environment bears no relationship whatsoever to the health risks they pose.

In 1973 the FDA proposed to make the law make sense. The agency recognized it had authority to specify the "method of analysis" that must be used to ensure "no residue." Why not first specify, for a particular carcinogenic animal drug, the safe level for humans, and then require that the drug's manufacturer develop analytical methods demonstrated to be capable of detecting levels at least as low as the safe level? If "no residue" of this drug were found in edible animal products using the method of analysis proved to be capable of detecting the maximum safe level, then the drug could be declared safe and legal. Here, the criterion for health protection becomes the controlling influence, and analytical chemistry has to be refined to ensure that that criterion is met. The problem was: how to define the health protective, or "safe level?"

At the time (early in the 1970s) the prevailing wisdom was captured in the convenient but somewhat misleading phrase "There are no safe levels of exposure to carcinogens." This phrase had been used by many experts on carcinogens in testimony offered to Congress on the occasion of its consideration of amendments to the basic food law, and in connection with other bills as well. Just what did this phrase mean?

Well, it was nothing more than a crude expression of the no-threshold hypothesis, as described in earlier chapters. Under this hypothesis, any exposure to a carcinogen increases the probability that cancer will occur. As we have emphasized several times, it does not mean that any exposure to a carcinogen will "cause cancer." If the exponents of the "no safe level" view meant that an absolutely safe level of exposure could not be identified, then they were correct, assuming the no-threshold hypothesis is correct. (Even assuming the threshold hypothesis is correct, as we have noted earlier, does not establish that we can ever be certain we have identified the completely safe threshold dose for any agent.)

But if the definition of safety were to be converted to one that was not absolutist, then the "no safe level" characterization no longer holds. The FDA took the position that the "safe level" for carcinogens such as DES was to be defined as that which produced no more than a specified, and very low level of excess risk. A risk assessment was to be carried out, based on animal carcinogenicity data obtained for the drug under consideration (the epidemiological data relating DES intake to human cancer, while clear, were not well suited for quantitative risk assessment). The low, "safe" risk level was to be specified; and the dose of the carcinogen corresponding to that "safe" risk level was to be estimated from the dose–response curve. Figure 8.1, as we have already seen, shows graphically how this is done; the "safe" risk level (on the vertical axis) is specified, and the corresponding "safe" dose is estimated from the linear, no-threshold, dose–response curve. The FDA proposed a one in one million lifetime risk level as the maximum allowable.

This approach was, in theory, more satisfactory than the absolutist approach, because it defined "safety" not in terms of the scientifically meaningless and indefinable "zero risk" standard (requiring banning, to ensure zero exposure, unacceptable as a general approach, as discussed earlier), but in terms that are scientifically meaningful because they do not require the impossible proof that something (risk) is absent. Safety, under this view, is a condition of very low risk.

Does this not pervert the meaning of "safe?" Perhaps. It might in principle be desirable to cling to the popular definition of safety – no risk – as a goal, but we also need to face facts. First, that no activity or exposure, no matter how "safe" it appears to be to the common sense, is demonstrably without risk. Whether we like it or not, we live with risk, it is unavoidable. There is nothing obviously wrong with seeking less and less risk (more and more safety), indeed many think there is a moral imperative to do so, but achieving zero risk is either not possible, or – and this is perhaps more relevant to the present discussion – is a condition that is not knowable. We shall note, but not further discuss, that seeking greater and greater reductions in risk in all circumstances is probably not wise public policy, given that we do not have infinite resources to spend on such reductions. Moreover, most risk reductions involve trading one risk for another; if this possibility is ignored, decision-making will be inadequate. But perhaps more important than any of this is the fact that the conduct of risk assessments in the context of what is almost always limited scientific understanding, and the attempt to use the results of risk assessments and often much other information in ways that satisfy the sometimes vague criteria set down in our laws, makes the development and enforcement of regulatory polices one of the most difficult undertakings that Congress has passed on to the Executive Branch. Hardly any decision that carries with it more than a modest financial burden goes unchallenged and many are subjects of enormous controversy.

While it is clear that some people will not accept a definition of safety that is relative, it appears that most people feel safe when they are convinced that risks to their well-being are sufficiently low, even if not completely absent. (There are some dramatic and important qualifications on this conclusion, as we shall see in the later section concerning people's perceptions of risk. While for the most part people accept that the condition of safety is not equivalent to the condition of being completely risk-free, most people do not perceive risk as simply a matter of probability, as do the experts. This intriguing and well-documented fact complicates greatly the public dialogue on matters of risk.)

## How safe is it?

While not everyone will be convinced that defining safety as a condition of very low risk is either wise or necessary, we shall proceed under the assumption that such a definition is the one that is most technically sound, and is the one that most people accept, either explicitly or implicitly. It is also the one that regulatory agencies have accepted, sometimes explicitly, sometimes not. The FDA, as we showed, initiated this approach to carcinogens, in 1973, and the EPA adopted it at about the same time.

In fact this same approach had been widely accepted, at least implicitly, prior to the FDA's actions. The methods long in use to establish ADIs for threshold agents cannot, as we have seen, guarantee absolute safety, even though they may appear to do so, and this conclusion holds for the more refined, but similar, method we have described for establishing RfDs and other health protective limits.

In any event, the FDA has declared that, for carcinogens such as DES, the condition of safety would be satisfied if the extra lifetime risk of cancer associated with consumption of residues in food did not exceed some very low level. In a regulatory proposal published in 1977, the agency spelled out some of the reasons it selected a one-in-one-million risk as that level. In essence, the FDA held that if this risk were accurate, and if every one of the 240 million people living in the United States at that time were exposed daily, for their full lifetime, to the residue concentration of carcinogen that created this risk, then the number of extra human cancer cases created over a 70-year average lifetime would be: $10^{-6}$ extra risk per person $\times$ 240 $\times$ $10^6$ persons = 240 extra cases per lifetime, or, an average of 240 ÷ 70 = 3 to 4 extra cases per year (for an average lifespan of 70 years). The agency then noted that the model used to estimate risk (the linear, no-threshold model) would be likely to overstate the size of that risk (for the same reasons set forth in Chapter 8); and, moreover, that it was very unlikely that anybody, let alone all 240 million citizens, would be exposed daily, for a full lifetime, to the maximum allowable food concentration of drug residue. So, the agency concluded that, the actual number of extra annual cancer cases associated with a one-in-one-million risk level, estimated using the regulatory risk model, would almost certainly be many fewer than 3–4, by an indeterminate amount. Given that there are nearly 1 500 000 new cancer cases per year in the United States (exclusive of skin cancers related to sunlight), it appears that the one-in-one-million risk level was an adequate definition of safety.

This was a policy, or *risk management*, choice on the part of the FDA, pursued to seek a method for limiting exposure to carcinogens that would rest upon the degree of risk posed, not the irrelevant capabilities of analytical chemists. The same, one-in-one million *insignificant* risk level would apply to all carcinogens used as drugs for food-producing animals, and this would result in variable allowable food residue levels, depending on the relative potencies of the carcinogens (lower allowable levels for more potent carcinogens).

## Expanding uses of risk assessment and the concept of insignificant risk

It became apparent to regulatory agencies during the 1970s that some means had to be developed to deal with carcinogens. Before that time carcinogens were either ignored, banned, or limited up to the amounts analytical chemists could detect. None of these was a very satisfactory approach to what was becoming obvious to all: more and more commercially important chemicals were being identified as human or, more typically, animal carcinogens; and chemists were finding these chemicals in more and more environmental media at lower and lower levels. Moreover, society was demanding, as evidenced by the spate of new laws passed in the United States in the 1968–1972 period, a clearer picture of the risks posed by these substances and greater control of those risks. At the same time the FDA was promoting the use of risk assessment/insignificant risk determinations for the very limited class of carcinogenic animal drugs, the scientific literature began to see more papers on risk assessment methodologies. All of this led the EPA to begin adopting risk assessment approaches for carcinogens, first in connection with pesticides, and eventually for all classes of regulated chemicals, including wastes found at Superfund and other such hazardous waste sites. The OSHA, which at first rejected risk assessment as a basis for regulating workplace exposures to carcinogens, eventually adopted the technique. Even the Consumer Product Safety Commission got in on the act, as it had to deal with carcinogens such as asbestos in certain hair dryers and formaldehyde in some home insulating materials. Many state regulatory authorities and regulators in the European Union and many other areas took up the mantle. Public health agencies are proceeding in similar fashion. All of this activity led to the development of the National Academy's 1983 "Red Book" which, we have seen, gave further impetus to the use of risk assessment.

There now exists a decision-making process for managing risks from carcinogens in the environment that includes the use of risk assessment, and the further notion that human health can in most cases be adequately protected by ensuring that risks do not exceed certain low levels. Moreover, regulatory policy makers have emphasized that no single risk level, such as the FDA's original one-in-one million level, satisfies the requirements of all the different laws that pertain to environmental carcinogens. Decisions about the appropriate risk goals for air pollutants, water pollutants, pesticide residues on food, food additives, occupational carcinogens, and so on, depend upon the requirements of applicable law, and policy makers have the responsibility to select risk reduction goals using the criteria set forth in these laws. Thus, as noted in the opening sections of the chapter, some statutes require zero risk for carcinogens (i.e., a complete ban on use, as required for certain food additives covered by one section, called the Delaney clause, of the Federal Food, Drug and Cosmetic Act, but not required for other classes of food chemicals). Others require that risk not exceed some specified, insignificant level; and under other laws, agencies are permitted to consider the technical feasibility of various risk reduction techniques, and, under yet others, some type of balancing of risks and benefits is permitted.

It is no wonder – and a source of confusion – that people can be exposed to different levels of risk from the same, regulated chemical depending upon whether they breathe it, drink it, consume it as part of their diet, or come into contact with it in their place of work or through the use of consumer products.

Moreover, under current laws, there is little opportunity for examining the totality of exposure to a single substance from all media, which is necessary if we are to acquire a picture of its total risk, and identify those sources that are the greatest contributors to that total risk. Risk decisions for chemicals in specific media are sometimes taken in isolation from decisions about the same chemicals in other media, and this practice is encouraged because of the diverse requirements of our laws and the artificial barriers they create.

But while recognizing these limitations in our laws and the imperfections of our regulatory institutions, let us now move on to the specific risk management approaches that are applied to the many sources of risk. As mentioned earlier, these brief reviews are of limited scope and emphasize only the various ways in which risk assessment results figure in risk management decisions.

## Food additives and contaminants

Except for the zero-risk case of chemicals (intentional, directly introduced food and color additives) covered by the Delaney clause of the federal food laws, which requires banning, carcinogens in food are controlled at different non-zero, risk levels. The FDA, which enforces the Delaney clause, has taken the position that certain classes of food ingredients not covered by that clause, can be permitted as long as excess cancer risks do not exceed some very low (i.e., insignificant) level. Drug residues in animal products are, as we have seen, among the classes of food chemicals permitted under this regulatory model. The FDA has also applied risk assessments to certain carcinogenic food contaminants (PCBs in fish, aflatoxins in peanuts and other products) and concluded that risks greater than the one in one million level normally applied to added substances can be tolerated, because these contaminants are not completely avoidable. For these substances the FDA applied a "balancing" approach: the risk associated with the contaminants balanced against the amount of contaminated food that would have to be removed from commerce if different tolerated levels of the contaminants were enforced. Aflatoxins and PCBs cannot simply be removed from food; foods containing excessive amounts of them have to be destroyed.

For the important case of methylmercury contamination of fish, the FDA has not established quantitative standards to force exposures below the RfD. Rather, the agency chose to issue warnings and fish consumption advisories to women who might be pregnant. The agency also noted the nutritional benefits for children of fish consumption by pregnant women, and sought to strike some balance between excessive methylmercury exposure and insufficient fish oil intake. Fish advisories of this sort have been used by many states to deal with contaminants arising in rivers and other water bodies under their jurisdiction. It has not been easy to determine whether such risk management approaches are truly effective.

The FDA also regulates food additives – substances, such as antioxidants, emulsifiers and non-nutritive sweeteners, that are intentionally and directly added to food to achieve some desired technical quality in the food. As noted, the Delaney clause prohibits the deliberate addition to food of any amount of a carcinogen. These additives, if they are threshold agents (not carcinogenic), can be allowed as long as the human intake does not exceed a well-documented ADI. Those who would seek approval for an additive need to supply the FDA with all of the toxicity information needed to establish a reliable ADI, and all of the product-use data that would permit the agency to assure itself that the ADI will not be exceeded when the additive is used.

## Pharmaceuticals

The risks of pharmaceutical agents are managed in quite a different way. Once it is determined that, through the conduct of pre-clinical and clinical studies (Chapter 8), a drug's risks and its benefits (its efficacy for reducing the risks of the medical condition it is to be used to treat) have been adequately documented by the drug's manufacturer, the FDA, often with the assistance of advisory boards composed of specialists, must decide whether the risks are outweighed by the benefits conferred. If such a conclusion can be supported, the drug can be approved ("licensed") and can be marketed.

A principal risk management tool for a drug is its so-called "labeling." With every drug the FDA issues detailed information on everything from its pharmacology and toxicology, to the results of clinical trials. Included is information on interactions with other drugs or with foods that may decrease safety or efficacy, warnings about side effects, and information on contraindications (patient conditions that indicate the drug should not be used). Of course the labeling also instructs on proper use and on appropriate dose rates.

Physicians are the prime recipients of this information, and their treatment of individual patients should be guided by it. (Guided is the correct word. The FDA cannot regulate individual physician practice. If information in the published literature supports use of a drug for conditions not described in the approved labeling, a physician has the right to prescribe for these "off-label" uses. One hopes there is evidence to support this; indeed, medical professional societies often develop position papers when evidence becomes available to support such uses.)

As everyone knows, the post-approval phase of a drug's life can bring news of significant new side effects. Post-marketing surveillance may turn up evidence of a type that cannot be uncovered in the typical clinical trials conducted in the pre-approval period. Rare medical conditions, or conditions that have latency periods longer than the duration of typical clinical trials, are difficult to detect. One type of post-marketing surveillance, wherein patients report to their doctors some adverse event, and doctors report that event to the FDA or to the drug's manufacturer (who has a legal obligation to report such adverse events to the FDA) may be useful, but these types of "adverse drug report" are nothing but fairly crude case-reports and are plagued by all of the difficulties we described in Chapter 6. The FDA seems to be ready to toughen its policies regarding post-marketing surveillance with requirements for so-called pharmacoepidemiology studies. Because of difficulties with certain painkillers that were uncovered in the year 2004, the FDA and the industry regulators are now under significant pressure to improve drug safety monitoring.

Drug labeling is an especially sophisticated form of information and warning, but there are similar risk management approaches in place for other products. Rather than prohibiting sales of the non-caloric sweetener aspartame, for example, because it is known that a relatively small number of people who suffer from phenylketonurea (PKU) could suffer serious side effects from ingesting the phenylalanine amino acid present in the compound, the FDA concluded that a warning directed at phenylketonurics against the use of products containing the sweetener provided adequate protection. Phenylketonurea sufferers know who they are (it is due to a genetic deficiency and all newborns are tested for it), and are trained from birth to avoid all phenylalanine-rich foods. Over-the-counter medicines and many cosmetics and household products carry warnings regarding the possible adverse consequences of improper product use. All over the country warnings regarding the consumption of fish can be found, to discourage individuals from incurring significant exposure to common fish contaminants such as PCBs, dioxins, and methylmercury. In California, Proposition 65, passed by popular vote in 1986, requires warning labels on products "known to the State to contain certain carcinogens or reproductive toxins." The Proposition does not require such warnings if product risks can be demonstrated to be sufficiently low, but severe penalties accrue to manufacturers who do not warn and who have not documented low risks. Nutrition labels, while not warnings, provide information useful for individuals to determine how to avoid the risks of inadequate nutrient intakes, and may soon contain similar information regarding the risks of excessive intakes. All of these types of information and warnings are risk management tools.

## Pesticide chemicals

Until 1994 the EPA regulated pesticides proposed for use on food crops under certain sections of the Food, Drug, and Cosmetics Act. Carcinogenic pesticides were subject to the Delaney clause, and were thus prohibited. The use of a non-carcinogenic pesticide was allowed if its manufacturer provided data sufficient to establish an RfD, and information on expected food residue levels sufficient to document that the RfD would not be exceeded when people consumed food containing residues of the pesticide. The tool for determining compliance with this criterion is called a *tolerance*, and it is expressed as the maximum amount of a pesticide that can be present in a given amount of food, if the RfD is not to be exceeded.

Assume, for example, that the RfD for pesticide Q is 0.2 mg/(kg b.w. day). This "safe" level will, for an average lifetime human body weight of 70 kg, allow a daily pesticide Q intake of 14 mg.

Assume further that the only use of pesticide Q is to treat certain insect pests on apples that are to be consumed as fresh fruit (no juice). Our expert on human food consumption patterns and rates tells us that "high-end" consumers of apples, those consuming at the 95th percentile of the distribution of consumption rates, eat approximately 0.3 kilograms (300 grams) each day (about two-thirds of a pound). If these consumers are not to ingest pesticide Q intakes in excess of

14 mg each day, then the apples they eat must not contain residues exceeding 14 mg/0.3 kg ≈ 47 mg/kg (ppm). The EPA will establish this level as an *official tolerance* for pesticide Q on apples. Those manufacturers who seek to market pesticide Q must demonstrate to the EPA that there is an analytical method capable of reliably measuring this level, and that apples treated in a specified way with pesticide Q do not contain residues in excess of this tolerance. The tolerance becomes a standard that must be met. If pesticide Q is permitted to enter the market, the enforcement agency, the FDA, will periodically sample apples to determine whether apples in commerce comply with the standard. If violations of the tolerance are found the agency will take action of some type to prevent sale and human consumption. Once a tolerance is established by a regulatory agency, it is published in the Code of Federal Regulations, and it becomes legally enforceable. The agency, to take legal action, does not have to demonstrate that a health problem exists, but only that a legal standard has been violated.

In the mid 1990s, following a report and recommendations from the National Academy of Sciences, Congress enacted and President Clinton signed the Food Quality Protection Act (FQPA), which introduced some important changes in the way pesticides were to be regulated. The primary impetus for change was growing concern, expressed in the Academy's report, about the possibility that children might be at increased risk from pesticides, especially during the developmental period of life. In particular, it was proposed that, because certain pesticides often acted through mechanisms damaging to the nervous systems of insects, the developing and fragile nervous systems of children might be especially vulnerable to such pesticides. The law also required the EPA to consider the aggregate exposures to all pesticides in a given class acting by the same toxic mechanism, and also to take on the very difficult task of regulating these aggregate exposures on the basis of the cumulative risks they pose. The EPA has attempted to deal with the large class of organophosphate cholinesterase inhibitors in this fashion, but progress is slow. But the FQPA contains an additional provision, intended to deal with the problems of increased childhood sensitivity and increased childhood exposures, that requires the EPA to divide a pesticide's RfD by a "safety factor" of up to 10. Such an explicit policy has not been encoded in other laws dealing with the regulation of chemicals. Interestingly, the FQPA drops the Delaney clause requirement for pesticides, and substitutes a one-in-one million lifetime risk criterion.

Under the EPA's interpretation of the Federal Fungicide, Rodenticide, and Insecticide Act (FIFRA), which applies to non-food uses and to certain uses in raw agricultural products, the agency is allowed to balance the risks associated with use of a pesticide against the benefits that would be lost were the pesticide not available. For pesticides found to be carcinogenic, the agency has tended to use the one-in-one million lifetime risk standard, but departs from it to allow somewhat higher risks when benefits are judged high, and seeks somewhat lower risks when benefits are thought to be negligible. Interestingly the EPA tolerates higher risks for exposures to pesticides incurred by workers who manufacture, distribute, or apply pesticides than they do for the general population. ("Balancing" is, if anything, much cruder than risk assessment itself; rigorous methods for measuring pesticide benefits and balancing against health and environment risks are pretty much unexplored, yet FIFRA requires it.)

## Occupational exposures

The toleration of higher levels of exposure for people who are exposed to chemicals on the job is not confined to pesticides. There is, in fact, a long tradition in toxicology to apply smaller uncertainty factors (for threshold agents) when establishing protective exposure levels for workers, than when establishing such levels for the general population, a topic we explored in Chapter 8. This makes sense in general, because workers are, on average, healthier than the general population, and the workforce does not include children, the infirm and the aged, and contains lesser numbers of individuals likely to be especially sensitive to chemical toxicity. Variation in susceptibility in a worker population is likely to be less than that in the general population. And, regulations compel the delivery to workers of extensive information about the chemicals they work with. In many cases their environments are monitored and, for workers handling especially hazardous chemicals, medical surveillance to detect early signs of a problem may be required. The Occupational Safety and Health Administration (OSHA), an arm of the Department of Labor, is the federal regulatory authority in these matters. The OSHA gets much of its scientific advice from the National Institute of Occupational Safety and Health. The "threshold limit values" (TLV®) published by the American Conference of Governmental Industrial Hygienists (ACGIH), while not official, have high standing in the field of occupational health, and are widely used. At present the ACGIH lists TLVs for many more chemicals than the number for which the OSHA has established PELs.

The OSHA has completed a number of rules on occupational carcinogens, including arsenic, benzene, asbestos, ethylene oxide and acrylonitrile. The agency conducted risk assessments and concluded that occupational exposure standards were too high and had to be reduced. A Supreme Court ruling in 1981 on the OSHA's first attempt to regulate a carcinogen (benzene) required the agency to adopt a risk-based approach, an approach the agency had first rejected.

The risks the OSHA estimated were based on the assumption that a worker could be exposed to the chemical for a working lifetime of 40–45 years, and that exposure each day of that period would be the maximum level permitted, the PEL. Because these exposure conditions are unlikely to exist for any individual, actual job-related risks are almost certainly lower than the levels the OSHA estimated, by unknown and varying degrees. Nevertheless, the excess cancer risks that the OSHA found tolerable, in most cases because of the technical limitations on achieving lower exposure levels, are greater than any that the EPA or FDA has seen fit to tolerate for members of the general population. While some FDA and EPA risk decisions on carcinogens go as high as one in 10 000, most are at lower risk levels; OSHA decisions on occupational carcinogens have generally not forced lifetime cancer risks below the one in 10 000 level, and some are higher.

It is hard to find compelling reasons to support the proposition that the workforce is less susceptible to cancer (as opposed to certain other forms of toxicity) than is the general population, so justification of the apparent "double standard" on these grounds is problematic. One of the OSHA's considerations in reaching decisions about tolerable risk levels has been information on job-related risks of other types. The agency has cited Bureau of Labor Statistics data on job-related fatalities arising from accidents and other hazards unrelated to chemical carcinogenicity. The OSHA found that lifetime risks of death associated with jobs most people perceive to be safe (office work, for example, or work in retail establishments) fall in the range of 1 per 1000 to 1 per 10 000, for a 40-year work period. In fact, the average lifetime risk of work-related death is 2.9 per 1000, in private sector establishments in the United States with more than 10 employees. Work-related risks of death in construction, mining, lumbering, and agriculture are 3–10 times higher. This type of information was used by the OSHA, together with arguments about technical feasibility, to support their decisions on occupational carcinogens. Note also that the OSHA requires some type of medical surveillance in cases where early detection of a developing cancer is possible. Workers on certain jobs are required to receive information and training to minimize their exposures.

# Drinking water

The EPA's Drinking Water Office sets limits on contaminants of drinking water, under the requirements of the Safe Drinking Water Act. This arm of the EPA establishes RfDs for chemicals that do not appear to be carcinogens, and then drinking water limits are set so that the RfD – actually a fraction of the RfD – is not exceeded. The use of only a fraction of the RfD allows for exposures to the same chemical through sources other than drinking water without the risk of exceeding the RfD. But for carcinogens, say the drinking water regulators, the goal for exposure ought to be zero. Because this ideal, called a maximum contaminant level goal (MCLG), cannot be achieved, enforceable standards for carcinogens – maximum contaminant levels – are established at the lowest technically feasible level. These levels typically translate to lifetime risks of one in 100 000 or lower, but for a few agents, arsenic being the most notable, risks associated with the maximum contaminant level are greater.

# Hazardous waste sites

The EPA makes decisions about clean-up of abandoned hazardous waste sites under the so-called "Superfund" law. Risk assessment outcomes are one guide to the decision process. The agency has declared that, for carcinogenic contaminants, clean-up must reach lifetime risks somewhere in the range of one in 10 000 to one-in-one million; most decisions seem to aim at risks of one in 100 000 or lower. Hazard index values for non-carcinogens are not expected to exceed one. Costs and technical feasibility figure heavily in these decisions.

The Resource Conservation and Recovery Act (RCRA), enacted in 1976, provides the principal authority to the EPA to regulate the handling and disposal of hazardous wastes. The many regulations now in place are directed at the intention to "protect human health and the environment" by law, and various direct and indirect measures of toxicity and possible human and environmental exposures guide regulatory requirements; risk-based criteria are thus a significant component of RCRA regulations.

## Industrial chemicals

Regulatory agencies possess, to varying degrees, the authority to require manufacturers to submit risk-related studies and information to the agencies when those manufacturers develop or otherwise uncover such information. For certain products – pesticides, food additives, pharmaceuticals – the manufacturers have, for many years, accepted the burden of proof of demonstrating safety (or safety and efficacy) before being allowed to market products, with the FDA and the EPA specifying the types of studies and data that must be developed, and the safety and efficacy criteria that must be met.

The Toxic Substance Control Act (TSCA), passed by Congress in 1975 and signed by President Carter in 1976, is implemented by the EPA. It contains provisions for industry reporting of new adverse health findings and significant new uses of existing chemicals, and toxicity testing requirements for new chemicals (those not present on an "inventory" the agency has developed of existing chemicals). Under TSCA the EPA can also require toxicity testing of existing chemicals, but does have the burden of establishing that there are sound reasons to require the data. The TSCA is concerned with industrial chemicals not regulated under FDA or other EPA-enforced laws.

If one surveys EPA and FDA regulatory documents one will turn up large numbers of protocols for the toxicity studies that these agencies require. In an effort to ensure uniformity in the data collected, these protocols have gradually achieved a high degree of standardization and efforts to ensure international standardization have been pursued intensely for the past two decades. This book's reference list provides guidance in locating these many toxicity testing protocols. The agencies have also issued Good Laboratory Practices regulations that require adherence to very high standards of scientific record-keeping during the conduct of these studies, and in results reporting. The GLP regulations came along after it was discovered, during the 1970s, that certain contract toxicology labs had "fudged" data. I recall, during my FDA career, receiving one particularly egregious example of such a study. Test Animal 4741 was described in the back-up data to the report as having died on day 332 of the study being reported. I noticed that the same animal was noted as having been weighed on day 336 and having the same weight as on day 315. Similar discrepancies abounded. Upon further investigation this was determined not just to represent sloppy record keeping; it was sloppy but genuine fraud. Lab directors were prosecuted. The GLP regulations were written.

## Air pollutants

We discussed regulation of the so-called "primary" air pollutants (ozone, $SO_2$, $NOx$, CO, PM, and lead) back in Chapter 4, along with current efforts directed at particulate matter. The Clean Air Act of 1991 provided the EPA with new authority regarding air pollutants. Because laws differ we find that in many important cases the government has the burden of demonstrating that a significant risk to health exists before action can be taken to institute management controls. The laws that place such burdens are generally those that deal with pollutants or unintended contaminants. Thus, in setting standards for the primary air pollutants under the original, 1970 version of the Clean Air Act, the EPA was required – through its own research programs and those of its sister public health agencies, and through the sponsorship of academic research – to develop the epidemiology, toxicology, and the source and exposure data necessary to support risk assessments (Chapter 4). The 1990 amendments to the 1970 law provided the EPA with the authority to regulate 189 specific hazardous air pollutants (HAPs). The EPA is required to issue for these HAPs emissions reduction standards based upon what is called "maximum achievable control technology" (MACT). After MACTs are in place, the agency is required to assess risks and decide whether the MACT-based standards are health protective, with an "ample margin of safety." This margin is explicitly defined in law for carcinogens: the exposure must create a lifetime cancer risk no greater than one-in-one million. If the imposition of an MACT-based standard does not result in emissions that meet the "ample margin" criterion, further reductions in exposure are to be imposed. The EPA has not yet fully regulated the 189 HAPs.

## Enforcement

Standards are not likely to have much impact unless there are authorities with the power to enforce them. And, where standards have not been established, those same authorities must have available the scientific and regulatory resources to assess which particular findings of contamination present risks to health of sufficient magnitude and seriousness to support enforcement action. Enforcement actions regarding chemical products range between the extremes of banning, and the often very costly, product recall, to the issuance of consumer warnings. Manufacturing plant inspections for compliance with Good Manufacturing Practice and other such regulations are important risk management activities of the FSIS (which has a force of meat and poultry inspectors placed in every plant), the FDA, the OSHA, the EPA, and all State agencies. Sampling and analysis of environmental media to determine whether tolerances or other limits on chemical exposures are adhered to are at the core of most enforcement activities for contaminants.

Risk management as undertaken by federal regulatory agencies has been the principal focus in this chapter, but similar decision-making is fast becoming a component of corporate life. Many manufacturers and users of chemicals have mounted programs to gain a better understanding of the risks their products and wastes pose – to workers, to consumers, to individuals exposed to emissions to the environment – and to undertake their own management actions, even when not yet demanded by regulations. Careful and honest evaluations of risk and the recognition that something must be done if risks are found to be excessive are among the trademarks of environmentally enlightened corporations.

## European Union – REACHing

Regulatory systems and institutions in the European Union, and in most of the rest of the world, are not unlike those we have described for the United States. Risk-based decision-making of the type used in the United States has only recently begun to be adopted in the EU and elsewhere, and the future pace of its adoption is difficult to predict. It does seem, however, that a common approach to risk assessment may be achieved on a global basis within the next decade. Indeed, common protocols for toxicity testing are now available and appear to be universally accepted; it is a good guess that common approaches for using test data to assess risk will follow.

Perhaps the most significant recent development regarding the control of chemical substances is represented by legislation now working its way through the EU governing bodies. As it is now written, the legislation pertains to all chemicals manufactured in or imported into the EU. The legislation has an intent similar to the TSCA law in the United States, but its requirements seem to allow less flexibility than do the TSCA rules. Indeed, as now drafted, the legislation seems to require that manufacturers assemble or develop significant amounts of toxicology and other types of data on perhaps 30 000 or more "existing" chemicals. The proposed program is called REACH: Registration, Evaluation, Authorization and possible restriction of CHemicals.

Companies will be required to provide information on the identity and properties of a compound (including physical, chemical, toxicological, and ecotoxicological properties), the intended uses, the estimated human and environmental exposure, risks for humans and the environment, and proposals for risk management measures. Information requirements will depend largely on production volumes, but might be adjusted based on the intrinsic properties and conditions of use of individual substances.

REACH is an extraordinarily ambitious program. There are discussions underway regarding proposals to limit the numbers of chemicals to be subjected to these requirements. The potential for toxicological testing on a massive scale raises questions about the availability of facilities to carry out such tests, and runs counter to the objective of reducing the numbers of animals used for such purposes. The need to accomplish REACH objectives without the overuse of laboratory animals has promoted discussion and research regarding the use of alternative methods to collect the necessary data: tools such as in vitro tests and quantitative structure–activity relationships (QSARs) are being promoted, and this has led to substantial research efforts to test their predictive validity. Time will tell where all of this activity leads us.

## Explaining risk-based decisions

Risk managers are confronted with a host of fairly complex technical, legal, and social issues when making decisions about whether to restrict people's exposure to consumer products, drugs, food ingredients, and environmental chemicals, and about the degree of restriction that is necessary. When all of this complex analysis is done, however, the manager needs to be able to face the public and declare that the final decision will ensure that their health will be protected. The effective manager will have to explain why a particular risk level is adequately protective and why it should be accepted by the public. Whether it is a manufacturing plant manager explaining to members of a surrounding community why they need not fear the emissions from the facility, or a regulatory official explaining why certain pesticide residues on tomatoes or a new food additive are safe to consume, the issue is pretty much the same – the public needs to be assured that its health is not jeopardized. And the public's understanding of the technical and policy issues needs to be sufficiently sound so that they can engage in a dialogue regarding the adequacy of decision-making. Public participation in the process would seem crucial to the success of any decision affecting people, but success is doubtful without an enlightened public.

The risk manager needs to be both confident of the wisdom of a decision and to be able to articulate clearly why it was made. He or she needs to have a fairly good understanding of the risk assessment underlying the decision, most especially the uncertainties associated with it, and how they were handled by the risk assessor. Confidence in the risk assessment, but not a foolish overconfidence, is essential. The manager must be assured that the assessment represents the current state-of-the-art, and also needs to be able to explain why the current state-of-the-art, although imperfect, is the best that can be done. Some people may expect certainty from science, but most recognize that the quest for certainty is an illusion and they will be put off by over-confident statements about what is known about risk, or its absence. Statements such as "we are sure this stuff is perfectly safe," even if uttered by highly regarded scientists (assuming such persons would ever make such a statement) only inspire mistrust.

The effective risk manager also needs to be able to explain why particular risk goals were selected. Most people can be made to appreciate the impossibility of a risk-free environment, although no doubt there will always be some who refuse to accept this notion, at least for that part of the environment containing industrial products. At the same time people are not willing to have a risk imposed on them that they perceive as unacceptably high, and will challenge decisions that do not satisfy them in this regard. Here we enter an area of discourse that is problematical, to say the least.

Decision-makers have sometimes found presentations of *comparative risk information* a useful aid to the public discourse on risk acceptance. We referred in an earlier section, for example, to the OSHA's use of statistics on the risks of job-related accidents to support decisions on risk reduction goals for workplace carcinogens. The agency noted that lifetime risks of death from injuries suffered in what most people perceive to be safe occupations fall in the range of about 1 per 1000 to 1 per 10 000. Data of this type were helpful in explaining why the agency settled on carcinogen risk levels in this range as sufficiently low to provide a safe work environment.

Professor Richard Wilson of Harvard University and an associate, Edmund Crouch, among others, have devoted considerable effort to collecting and analyzing risk information on activities commonly engaged in and exposures commonly incurred. This type of information can be used effectively to both educate the public about risk in general and to assist risk managers' efforts to explain specific risk decisions.

Some of the risk data assembled by Wilson and Crouch are presented in Tables 11.2 and 11.3. In Table 11.2 are lifetime risks for a number of activities and exposures that most people undertake or experience.

Note that some of the risk information is actuarial (based on statistical data, typically collected and organized by insurance companies), and some of it has been derived from the type of risk assessment discussed in this book (chloroform in chlorinated drinking water, aflatoxin in peanut products). While the uncertainties associated with the figures in Table 11.2 are much greater for some risks than for others (not a trivial problem in presentation of risk data), such a presentation, it would seem, is helpful to people who are trying to acquire some understanding of extremely low probability events, of the order of one-in-one million.

One of Wilson's more interesting presentations is that depicted in Table 11.3. Here exposures or activities associated with annual (not lifetime) risks of one in 1 000 000 have been described, another useful way to help people gain some sense of the "reality" of very low probability events.

Comparative risk analysis is undoubtedly highly informative and can help risk managers to make decisions and then to explain them. But there is another issue here: it is quite clear from a good deal of research by social scientists that people's notions about risk are considerably more complex than those of the experts. *People do not perceive various threats to their health and well being simply as matters of probability.* Many attributes of a potential threat, besides its probability of occurring, influence people's judgments about whether they are willing to tolerate it, or, as Professor Peter Sandman of Rutgers University puts it, contribute to a determination of how much "outrage"

Table 11.2 *Annual risks of death associated with some activities and exposures, as compiled by Edmund Crouch and Richard Wilson*

| Activity/exposure | Annual risk (deaths per 100 000 persons at risk) |
|---|---|
| Motorcycling | 2000 |
| All causes, all ages | 1000 |
| Smoking (all causes) | 300 |
| Smoking (cancer) | 120 |
| Fire fighting | 80 |
| Hang gliding | 80 |
| Coal mining | 63 |
| Farming | 36 |
| Motor vehicles | 24 |
| Rodeo performer | 3 |
| Fires | 2.8 |
| Chlorinated drinking water (chemical by-products) | 0.8[a] |
| 4 tbsp peanut butter/day (aflatoxin) | 0.8[b] |
| 3 oz charcoal broiled steak/day (PAHs, Chapter 5) | 0.5 |
| Floods | 0.06 |
| Lightning | 0.05 |
| Hit by meteorite | 0.000 006 |

[a] Assumes water contains maximum level of by-product permitted by EPA; most water supplies contain less.
[b] Assumes aflatoxin present at maximum FDA-permitted level; most commercial brands contain much lower levels.
Source: Crouch and Wilson as cited by Slovic, P., 1986. Informing and educating the public about risk. *Risk Analysis*. 6, 403–415.
Note: Risks from activities are actuarial and much more certain than those associated with chemical exposures, which are estimated using regulatory models. Risks of cancer are assumed to equate to risks of death. Lifetime risk will be about 70 times higher if risks do not change substantially from year to year.

they feel. Woe to decision-makers who do not consider the implication of the work of investigators such as Sandman, and of Paul Slovic and colleagues at Decision Research, in Eugene, Oregon, and Baruch Fischoff at Carnegie-Mellon. This is both fascinating and important.

Investigators such as Slovic, Fischoff and Sandman have studied how people perceive, or feel about, potential threats. Some of the attributes of a particular risk people consider, either explicitly or implicitly, in forming judgments, are the degree to which it

Table 11.3 *Risks estimated by Wilson to increase chance of death in any year by 0.000 001 (1 chance in 1 million)*

| Activity | Type of risk |
|---|---|
| Smoking 1.4 cigarettes | Cancer, heart disease |
| Spending 1 hour in a coal mine | Black lung disease |
| Living 2 days in New York or Boston | Air pollution |
| Traveling 300 miles by car | Accident |
| Traveling 10 miles by bicycle | Accident |
| Flying 1000 miles by jet | Accident |
| Living 2 months in Denver on vacation from New York | Cancer caused by cosmic radiation |
| Living 2 months with cigarette smoker | Cancer, heart disease |
| One chest X-ray taken in a good hospital | Cancer caused by radiation |
| Eating 40 tbsp of peanut butter | Liver cancer caused by aflatoxin[a] |
| Drinking 30 12-oz cans of diet soda | Cancer caused by saccharin |
| Living 150 years within 20 miles of a nuclear power plant | Cancer caused by radiation |
| Risk of accident by living within 5 miles of nuclear reactor for 50 years | Cancer caused by radiation |

[a] Assumes aflatoxin at maximum FDA-permitted level; most commercial brands contain much lower levels.
Source: Wilson, R. 1979. Analyzing the risks of daily life. *Technology Review*. 81(4), 41–46.

is voluntarily assumed, the extent to which a personal benefit is perceived to exist as a result of incurring the risk, and the degree to which it is felt that there is personal control over the risk.

Generally, people are much more willing to tolerate risks that are voluntary than those they perceive as imposed upon them. It is not surprising, then, that when they learn about benzene emissions into the air they breathe, people living near a petroleum refinery are not going to be easily satisfied by an explanation that the lifetime risks of cancer associated with these emissions are no greater than one in 100 000, even though these estimates are probably pretty conservative (they overstate the actual risk), and even though they are more than 1000 times less than the voluntarily assumed risks of death from driving an automobile! Although people living in such circumstances are free to move away, most people would consider such exposures largely involuntary – they would only be truly voluntary if someone knew about them before moving to the affected area.

Degree of control is also important. The risks of riding in an airplane are perceived by many people to be much greater than the objective facts reveal. Part of the reason for this is that people tend to fear catastrophic events, such as a crash that may kill or injure many people at once, much more than they do events that take only one or a few lives at a time (such as accidents involving the very much riskier means of transportation, the automobile). But another part of the fear concerns the fact that people riding in airplanes feel they have absolutely no control over their fate. People driving automobiles feel safer than do passengers, for the same reason. Not rational? Perhaps, but nevertheless a common feature of human psychology.

Pesticides and food additives can provide many benefits, not only to food producers but also to consumers. But most people are not very aware of these benefits, or at least do not personalize them to a high degree, and this no doubt contributes to their sense of outrage when they hear about new health risks from these sources. Why should I take a risk when the only people deriving the benefits are the manufacturers? Some people take very high risks, whether on the job, as part of their recreational activities, or resulting from their personal habits (smoking, excessive alcohol consumption), because they feel they are getting something out of it for themselves; but they become upset when asked to tolerate very much smaller risks from activities or exposures that they feel are without significant personal benefits. Even if, in fact, they do derive some personal benefit, unless this is known to them, or is internalized in some way, people will not perceive it, and perception is what is important here.

Perception of personal benefit contributes to the pursuit of many highly risky "lifestyle" choices, and makes risk-management for public health authorities a difficult undertaking. Perception of the absence of personal benefit contributes to the strange resistance of individuals to the relatively small risks associated with involuntary exposure, and this makes risk management difficult for regulatory officials.

Slovic, Fischoff, and others have found numerous other attributes of a potential health threat to be important influences on people's perceptions. A threat that is of natural origin is more readily tolerated than one of industrial origin. Risks associated with familiar technologies are much less fearsome than those arising from new technologies (the products of biotechnology, for example, and perhaps those of nanotechnology). Some diseases or injuries are perceived to be far more dreadful than others (cancer is certainly near the top), and risks that create such conditions are accordingly more dreaded. Even this brief sketch should prompt the reader to want to learn more about this extraordinarily interesting topic, so more references to some primary works in the field are provided in *Sources and recommended reading*.

These features of human psychology (I am not a social psychologist so I shall not attempt to deal with the topic of why people perceive threats as they do) are important factors in the public discourse on risks from chemicals in the environment. It is not difficult to discern that risks from synthetic chemicals in the environment, whether they be contaminants or useful industrial products, tend to be among those for which people have the least tolerance. Exceptions might be products that have been around a long time and that have come to be seen as important in people's lives. The most pronounced expressions of outrage brought about by revelations during the 1970s that saccharin, the artificial sweetener used since the turn of the century, was an animal carcinogen, came from those who did *not* want the FDA to ban it (indeed, Congress passed a special law to keep it on the market!). But saccharin and products like it are exceptions to the general rule. So governmental and corporate officials who have to defend their risk reduction decisions regarding environmental chemicals have a doubly tough task. They not only have to be able to explain why particular risk levels are adequately protective – i.e., that absolute safety is not achievable and that it need not be achieved to protect the public health – but they also need to be able to deal with people's inherently low tolerance of these types of risk. This is really hard, perhaps even impossible to accomplish with anything close to perfection, but it must be attempted. How do we get people to worry about and act upon the major risks to their lives, especially those that they do not perceive as particularly threatening, and stop them worrying excessively about minor risks? This is a tall order, but very important, and the type of understanding created by the research of Slovic, Sandman and others is immensely helpful towards this effort. It is also helpful in preparing the way for wider public discourse on these topics, so that people can participate actively in decision-making.

Our review of some regulatory decisions, based on risk information, has been a relatively superficial one, and has avoided many complicated legal and policy issues, to say nothing of the political warfare that may accompany some decisions. The review, though selective in its coverage, does reveal that regulators draw no single line of demarcation to separate "safe" from "unsafe" exposures to chemicals – i.e., that safety is a relative condition, defined according to the degree of risk found tolerable under specific circumstances. We have also seen that regulatory authorities tolerate a fairly wide range of risks, based on varying legal requirements and historical precedents, for specific classes and sources of environmental chemicals; it is also apparent that these differences in risk toleration are difficult for the public to understand and, if truth be told, are not always easily understood by close observers of the regulatory scene. Public perceptions of risk clearly do not match well with those of the experts; education may serve to affect perceptions to a degree, but it is unlikely that the factors that affect people's fears will change in any fundamental way. Risk management decisions, except the most trivial, will always be difficult and unsatisfactory to some of those involved in or affected by them, and we can only expect to improve this state-of-affairs by ensuring that the best available scientific and technical information is incorporated, and that such decisions are made in the most open way possible.

# 12

# A look ahead

Before looking to the future, it will be useful to look back at some issues related to risk assessment and management that have so far been ignored. The issues are not so much technical as they are social and political, and to ignore them completely could leave the misleading impression that all the scientific and policy questions we have discussed are interesting matters for scholarly debate, and not much else. We shall not make this mistake.

Risk assessments reveal public health problems, of greater or lesser magnitude. If a problem is uncovered, we cannot simply hide it (at least not easily); we need to do something to reduce or eliminate it. Somebody will have to pay, no two ways about it. Depending upon the problem, costs could be massive for society as a whole, massive for selected industries, or, at the other extreme, relatively small all round. The latter generally raises only a little smoke, but when costs are heavy, things burst into flames. Because the industries that must bear the cost do not wish to be seen as destroyers of the public health or the environment, some tend to begin by determining whether there are credible ways to attack the scientific quality and accuracy of the risk assessments regulators are relying upon. They may claim risks have been exaggerated, that there is no, or only a minor public health problem. They may attack the regulator's risk management strategies, and claim there are less costly ways to achieve the same risk reduction goals, or that regulators are straying beyond their legal mandate.

On some of these risk questions affected industries, or their scientists, may have significant points to make, and may affect regulatory assessments if their arguments are persuasive. These debates may go on for very long periods of time (the EPA has been "re-evaluating" dioxin-related risks since 1992, and the work remains incomplete), and many make their way to the public, and arouse more concern, and sometimes anger.

Of course the consumer and environmental advocates that are now a major presence in most societies also have their say, and, as can be expected, often find the regulators deficient in their risk assessment science and negligent as risk managers. When major product or environmental controversies erupt, experts can always be found who will take positions at the extremes and who thereby leave the public more confused. Depending upon their personal political convictions, citizens tend to blame industry for obfuscating the truth, or blame government agencies for incompetence or for unnecessarily "crying wolf." We have also seen how varying perceptions of risk can strongly influence these debates, often more so than the more purely technical matters.

So many of the types of scientific and policy activities we have been attempting to explain, and their limitations, which we have not hidden, are frequently undertaken in the glare of the public spotlight, but with more heat being generated than light being cast. Scientists with no apparent leaning on these matters (note the word "apparent") other than getting the science right (not easily definable in the risk assessment arena) may end up dismayed, feeling that science is being ignored or distorted for political purposes. At the other extreme, an anti-science view – one based on the notion that, because science cannot tell us everything, we should ignore it altogether – seeks to influence public policy by invoking vague, subjective, decision-making criteria (e.g., "caution at any cost"). And somewhere out on another pole sits the view that the type of risk assessment approach we have been describing is bogus science, filled with wholly unwarranted assumptions and untested hypotheses, a naive view of science that would lead to virtually no basis for regulating exposures to most chemical substances (which is perhaps the desired outcome for those advocating this extreme position).

While these contending forces will no doubt continue to contend, and to confuse the public, we should hope that mainstream science and thoughtful risk management policies will prevail. The risk assessment framework is central in the continuing search for the right balance point in these debates.

## A taxonomy of risk

The risk assessment framework, first proposed in 1983 by the National Academies committee that produced the Red Book, has proved to be durable. Its influence has extended to other areas of risk assessment, and public health authorities who are responsible for understanding and mitigating the truly large public health problems of our time, those associated in part with certain personal behaviors, are beginning to describe their activities within that same framework.

In the broadest possible sense, risk assessment might be described as the analytical framework within which all public health problems of environmental origin are evaluated, and risk management is the analytical framework used to evaluate and decide upon strategies to reduce risks found to be excessive. Risk assessment is thus the framework used to track progress in efforts to reduce public health risks of environmental origin. In the sense used here, risks of environmental origin include those arising both by personal choice as well as those incurred involuntarily. Risk management strategies, as described in Chapter 11, must be tailored to the nature of the risk under scrutiny.

Adoption of this all-encompassing definition by public health and regulatory institutions could lead to a greater and more coherent understanding of the health threats that face us. We would see, in as systematic a way as possible, and with attention to the uncertainties in our knowledge, that our populations are faced with a relatively small number of very large health threats, a sizeable number of moderately large risks, and a very large number of relatively small risks. The very large risks are typically those that are called "lifestyle" in origin, and include smoking, dietary and alcohol abuse, and a host of safety related problems that lead to accidents. These risks are, of course, not only related to "lifestyle choices," but are also caused by safety problems due to manufacturer or other error outside the victim's control. Indeed, smoking has come to be seen not only as a lifestyle choice, but also a source of risk imposed by tobacco companies through the delivery of addictive nicotine. So, while the term "lifestyle" choices is commonly used by public health officials, its unqualified use may overly blame the victim.

These risks are, for the most part, large ones, and can be measured directly, by epidemiological studies and also by the activities of actuaries, who find ways to collect statistical data on rates of accidents and other safety risks, and on how they are distributed in populations. All-in-all, the relatively few risks that are large – the major causes of illness and death – are directly measurable and so are known with a relatively high degree of accuracy.

Managing these "lifestyle" risks proves to be extraordinarily difficult. Behaviors must be changed and, of course, they cannot be regulated in anything like the same way the types of exposures that have been the principal subjects of this book can be regulated. So risk management becomes education, nutrition and exercise advice, warnings, and, where possible, a degree of regulation (seat-belt laws, for example, and restrictions on smoking to limit the risks from side-stream constituents). It is no secret that getting people to change harmful behaviors requires no small effort.

The substances we have described in this book turn out to be, for the most part, minor threats to our health. There are, however, perhaps thousands of these threats, and so a regulatory system has evolved to reduce the risks of individual substances to very low levels. There is considerable work yet to be done, to learn the toxic properties of many poorly tested chemicals, and to assess their health risks, but once this is done, regulatory officials have available strong weapons (which they do not always choose to deploy) to reduce exposures when necessary. The burden for exposure reduction falls not on exposed individuals (as in the case of lifestyle factors), but rather upon those who have caused the exposures to occur.

Between the relatively small number of large risks and the relatively large number of small risks, we find an intermediate number of moderately sized risks. Many are occupational. Some are related to the use of pharmaceuticals and various medical products. And some are of environmental origin. Some of the leading causes of air pollution may be in this last category. Some of these risks are large enough to be detected and measured by epidemiologists, but others can only be estimated using risk assessment methods. Food and Drug Administration oversight, product warnings and individual medical guidance are the usual risk management tools for medical products, which, like any risk management system, can sometimes fail. Occupational risks are subject to control through the institution of good industrial hygiene practices and monitoring of work place environments and sometimes workers themselves. The more significant general population risks have, at least in developed countries, declined significantly in the past several decades, but some remain and more will be found. Air pollution risks are particularly troublesome, and also turn out to be among the most costly to control. We should also not ignore the significant risks associated with food-borne and water-borne pathogens introduced in Chapter 9.

Pressures on regulators to reduce the remaining significant risks will continue, as will pressures to force the collection of greater volumes of data on poorly tested but widespread chemicals.

The description of risks just given – a kind of taxonomy of risk – would take on a different cast if we were to move to the lesser developed countries. The sources of the major causes of morbidity and mortality in the lesser developed world have much more to do with infectious disease, food-borne and water-borne pathogens, and with significant undernutrition. Many of these conditions take large tolls, and they are not related to "lifestyles" (of course smoking, alcohol abuse, and overconsumption of foods are also significant problems, especially among the wealthier segments of these societies). Lack of food and basic sanitary systems are large issues. At the same time, as much basic industry moves into the lesser developed countries, the chemical problems we have described here will spread around the world – indeed, they already have, with perhaps China in the lead.

There is a significant degree of worldwide movement to adopt the risk assessment framework we have been describing as a public health model, one that can help us to understand public health risks and to monitor progress in reducing them, in a highly systematic way, and in a way that allows for relatively uniform measures of that progress. Indeed, a number of researchers in the field have been evaluating various uniform measures of the effects of different diseases. Some work, for example, is aimed at developing single "quality of life" measures that can be applied to all causes of morbidity and mortality. The work is still controversial and not widely accepted or understood, but further experience may prove the method to be convincing and useful.

## The value of risk assessment

There are occasional outbursts of skepticism about the value of the risk assessment approach. Some argue its full use is overly burdensome, that the only real need is to identify toxic chemicals and then eliminate them or reduce their presence. Perhaps this approach is appropriate for a few chemicals that exhibit extremely high toxicity or that are excessively persistent and bioaccumulate in an environment, and for which there are reasonably good substitutes available. But as a general approach, it ignores the elementary principle that all chemicals will exhibit toxicity at some dose. A regulatory or public health approach based on simply eliminating "toxics" is no approach at all. Moreover, if a program to reduce exposures to any "toxic" to the maximum extent possible, two opposite and detrimental outcomes can be envisioned: (1) the reduction may put an end to the use of a beneficial chemical without any demonstration that it is harmful (risky) under its conditions of use; and (2) without a risk assessment, it is not possible to claim that the exposure reduction achieved will be sufficient to avoid excessive risks (the "maximum reduction" may not be good enough).

Those who attack the use of risk assessment are sometimes confusing the assessment with the scientific data upon which it is based. Risk assessment is not scientific research; it cannot produce new data, it cannot create new knowledge. Rather, risk assessment, if it is well done, allows us to see clearly what we can claim to know and what we do not know about a given source of risk, based on whatever scientific information we have been able to acquire. Risk assessment should not be blamed for society's failures to acquire adequate scientific data and knowledge. Indeed, a completed risk assessment, incorporating all available data and knowledge, reveals most clearly the types of research that will lead to a reduction in whatever uncertainties remain. Use of risk assessment results as a highly systematic guide to research was an important insight of the Red Book committee, but it has received little attention.

Risk assessment is often confused with "risk–benefit" analysis, or "risk–cost" analysis, terms that some individuals who are strong defenders of public health find, if not offensive, then at least irrelevant. But risk assessment is not equivalent to "risk–benefit" or "risk–cost" analysis. Risk assessments yield information about risks to public health from many sources. They can also tell us what kinds of health benefits (risk reductions) can be expected if various actions are taken or regulations imposed. Some of our laws require regulatory officials to evaluate the relative economic costs of management options, or to consider the possibility that certain benefits might be lost if risk reduction actions were taken. These are separate activities, and they are not inherent features of risk assessment. Indeed, in a few cases – food additives come to mind – no offsetting benefit or cost calculus can be used; their approval by the FDA rests entirely upon a finding of insignificant risk ("reasonable certainty of no harm").

Winston Churchill said that "democracy was the worst form of government, except for all the others." I paraphrase: "Risk assessment is the worst basis for public health decision making, except for all the others."

## More data, more understanding

Two future trends have been hinted at throughout this book. One trend concerns the societal pressures to acquire far more complete data regarding the adverse health effects of ever greater numbers of chemicals and on the extent of human exposure to them. Acquiring such information, it is said, will pave the way for eliminating or reducing exposures to as many industrially produced chemicals as possible, and so provide improved public health. A second force, espoused by many basic scientists, urges greater attention on research, toward the goals of improved understanding of toxic phenomena, closing the gap between experimental and epidemiological approaches to gaining that understanding, and, in the end, to improve the basis for assessing health risks. In some ways these are complementary forces, but they do lead to competition for limited resources between the advocates of toxicity and exposure data gathering and those who seek to understand more completely the data we gather. One might say that the pressure to engage in more data gathering studies – whether experimental or epidemiological – comes primarily from the regulators and environmental and consumer advocates, while the drive toward mechanistic investigations finds its principal impetus among manufacturers, and from the research community. As we have suggested at several points along the way, many believe that greater understanding, through mechanistic investigations, will reveal that the typical government-style, default-driven assessments will lead to much risk overestimation, and that it provides a distorted picture of the extent of the public health problem chemicals pose. Although both sides of this issue no doubt acknowledge the value of the other's viewpoint, it is likely that these tensions will remain for some time.

The substantial growth of interest in toxicology, and of epidemiological investigations of chemical toxicity, has now brought many basic scientists into the disciplines. They are attracted no doubt in part because of the intense public focus on these issues, but perhaps more so because of the extraordinarily interesting scientific challenges associated with the effort to understand health risks of environmental origin. Much of the work is closely linked to advances in basic biological phenomena. Moreover, there are few subjects that require collaborations among so many scientific disciplines of both applied and basic nature. As the sciences necessary to support public health risk assessments mature, they gain standing within the larger scientific enterprise, and so have the potential to become significant components of the science education programs of our schools and colleges. Education in the matters that are the subjects of this book is perhaps the path necessary to reducing some of the rancorous discourse and confusion that so often accompanies public health decision-making. Given the pathetic state of scientific understanding among the general population, as evidenced in part (but not only) by the so-called debate pitting "creationism" (or its more sophisticated version, "intelligent design") against the Darwinian view of life, we should not expect rapid progress in building an informed public, but the efforts to do so should not falter.

## A definition for the future?

Looking forward with hope to the continued development and use of the risk assessment framework, and of those scientific disciplines that supply the data and knowledge necessary for its use, I propose the following, perhaps overly grand, definition of risk assessment, in an attempt to ensure its applicability not only to the types of problems that have been the principal subjects of this book, but to all those of public health concern.

Risk assessment is the analytical framework used to organize, evaluate, and characterize available knowledge and its associated uncertainties regarding the nature and magnitude of threats to human health arising from the environment, including both the natural world and every type of human influence on it. The results of risk assessments are used to guide policy decisions regarding the need to take actions to control or eliminate these threats so that human health is adequately protected. They are also used to identify the research needed to reduce uncertainties and, thereby, to improve understanding of these threats. Risk assessment is thus the instrument used to measure progress in understanding and managing every type of environmental threat to human health.

The definition could equally apply to threats to our environment, which may, in the long run, be the most significant determinants of the health of our planet and of all of its inhabitants.

# Sources and recommended reading

As in most areas of science, the information available on the subjects of this book, and the ease with which it can be acquired, have improved immensely since the first edition was published. Huge amounts of useful and reliable information are, of course, easily retrievable from the Internet, and a number of new and comprehensive works in toxicology and risk analysis have been published in the past decade. I list below the principal sources of information for the book, all of which can be recommended to those who might wish to pursue further any of the topics I have covered.

## Some general reference works

Most of the subjects covered in this book are dealt with in the following multi-author volumes.

Derelanko, Michael ed. (1995) *Handbook of Toxicology*, 2nd edn. Boca Raton, FL, CRC Press.
  Available from the same publisher as a "pocketbook."
Hayes, A. Wallace ed. (2001) *Principles and Methods of Toxicology*, 4th edn. Philadelphia, Taylor and Francis.
Klassen, C. D. ed. (2001) *Casarett and Doull's Toxicology: The Basic Science of Poisons*, 6th edn. New York, McGraw Hill.

## Chapters 1, 2, and 3

Centers for Disease Control. (2001) *National Report on Human Exposures to Environmental Chemicals*. Washington, DC, Department of Health and Human Services.

Gilbert, S. G. (2004) *A Small Dose of Toxicology*. Boca Raton, FL, CRC Press.

International Programme on Chemical Safety. (1999) *Principles for the Assessment of Risks to Human Health from Exposure to Chemicals*. Geneva, World Health Organization. Chapter 5.

Lippman, M. ed. (2000) *Environmental Toxicants: Human Exposures and Their Health Effects*, 2nd edn. New York, John Wiley and Sons.

Lüllmann, H. *et al.* (2000) *Color Atlas of Pharmacology*, 2nd edn. Stuttgart, Thieme.

Natural Research Council. (1993) *Pesticides in the Diets of Infants and Children*. Washington, DC, National Academies Press.

Smiley, R. A. and Jackson, H. L. (2002) *Chemistry and the Chemical Industry: A Practical Guide for Non-Chemists*. Boca Raton, FL, CRC Press.

USEPA. (1992) *Guidelines for Exposure Assessment*. Washington, DC, EPA. Publication No. EPA/600/Z-92/001.

Wexler, P. ed. (1998) *Encyclopedia of Toxicology*. San Diego, Academic Press.

## Chapter 4

Ballantyne, B., Marrs, T., and Syversen, T. eds. *General and Applied Toxicology*. (1999) 2nd edn. New York, Groves Dictionaries Inc., and London, Macmillan Reference Ltd.

International Programme on Chemical Safety. *Environmental Health Criteria*. Geneva, World Health Organization.

Multiple volumes on the toxicology of individual chemicals are available.

Lu, F. C. and Kacew, S. (2002) *Lu's Basic Toxicology*, 4th edn. New York, CRC Press.

Massaro, E. J. ed. (1997) *Handbook of Human Toxicology*. New York, CRC Press.

## Chapters 5 and 6

Ames, B. N. *et al.* (1973) Carcinogens are mutagens: a simple test system combining liver homogenates for activation and bacteria for detection. *Proceedings of the National Academy of Sciences*. 70, 2281–2285.

Armitage, P. and Doll, R. (1954) The age distribution of cancer and a multi-stage theory of carcinogenesis. *British Journal of Cancer*, 8, 1–12. Reprinted, with commentary, in: *International Journal of Epidemiology*. 33, 1179–1184 (2004).

Carson, R. (1962) *Silent Spring*. Boston, Houghton Mifflin.

Cohen, S. M. (2004) Human carcinogenic risk evaluation. An alternative approach to the two-year rodent bioassay. *Toxicological Sciences*. 80, 227–229.

Conklin, G. (1949) Cancer and environment. *Scientific American*, **180** (1).

Davis, D. L. *et al.* (1990) International trends in cancer mortality. *The Lancet.* **366**, 474–481.

Doll, R. (2004) Commentary: the age distribution of cancer and a multistage theory of carcinogenesis. *International Journal of Epidemiology.* **33**, 1183–1184.

Doll, R. and Hill, A. B. (1950) Smoking and carcinoma of the lung. Preliminary report. *British Medical Journal.* **2**, 739–748.

Doll, R. and Peto, R. (1981) The causes of cancer: quantitative estimates of avoidable risks of cancer in the United States today. *Journal of the National Cancer Institute.* **66**, 1191–1308.

Ellerman, V. and Bang, O. (1908) Experimental leukemia in chickens. *Zeitschrift fur Hygiene und Infektionskrakheiten.* **63**, 231–273.

Gordis, L. (2000) *Epidemiology*, 2nd edn. Philadelphia, W. B. Sanders.

Hueper, W. (1942) *Occupational Tumors and Allied Diseases.* Springfield, IL, C. C. Thomas.

Kennaway, E. L. and Hieger, I. (1930) Carcinogenic substances and their fluorescent spectra. *British Medical Journal.* **1**, 1044–1046.

McCann, E. *et al.* (1975) Detection of carcinogens as mutagens in the Salmonella/microsome test: assay of 300 chemicals. *Proceedings of the National Academy of Sciences.* **70**, 5135–5139.

Nasca, P. and Pastides, H. (2001) *Fundamentals of Cancer Epidemiology.* Gaithersburg, MD, Aspen Publishers.

Rous, Peyton (1911) A sarcoma of the fowl transmissible by an agent separable from the tumor cells. *Journal of Experimental Medicine.* **13**, 397–411.

Varmus, H. and Weinburg R. A. (1993) *Genes and the Biology of Cancer.* New York, Scientific American Library.

Williams, B. (1993) *Biostatistics: Concepts and Applications for Biologists.* London, Chapman and Hall.

Wynder, E. L. and Graham, E. A. (1950) Tobacco smoking as a possible etiologic factor in bronchogenic carcinoma. *Journal of the American Medical Association.* **143**, 329–336.

*Report on Carcinogens: Tenth Annual Report* (2001) Department of Health and Human Services, National Institute of Environmental Health Sciences. Research Triangle Park, North Carolina.

## Chapters 7 and 8

Haimes, Y. (1998) *Risk Modeling, Assessment, and Management.* New York, John Wiley and Sons.

Hardman, J. G. and Limbird, L. E. (2000) *Goodman and Gilman's The Pharmacological Basis of Therapeutics*, 10th edn. New York, McGraw Hill. Chapters 3 and 4.

Hrudey, S. and Krewski, D. (1995) Is there a safe level of exposure to a carcinogen? *Environmental Science and Technology.* **29**, 370A.

National Research Council. (1983) *Risk Assessment in the Federal Government: Managing the Process.* Committee on the Institutional Means of Assessment of Risks to Public Health. Washington, National Academy Press.

National Research Council. (1994) *Science and Judgment in Risk Assessment.* Committee on Risk Assessment of Hazardous Air Pollutants. Washington, National Academy Press.

National Research Council. (1996) *Understanding Risk: Informing Decisions In a Democratic Society.* Washington, National Academy Press.

Paustenbach, D. ed. (2002) *Human and Ecological Risk Assessment: Theory and Practice.* New York, John Wiley and Sons.

USEPA. See Internet section, below, for links to all EPA guidelines on risk assessment.

## Chapter 9

Butenhoff, J. L., *et al.* (2004) Characterization of risk for general population exposure to perfluorooctanoate. *Regulatory Toxicology and Pharmacology.* **36**, 363–380.

Caitley, R. C. *et al.* (1998) Do peroxisome proliferating compounds pose a hepatocarcinogenic hazard to humans? *Regulatory Toxicology and Pharmacology.* **27**, 47–60.

Calabrese, E. J. and Cook, R. R. (2005) Hormesis: how it could affect the risk assessment process. *Human and Experimental Toxicology.* **24**, 365–270.

Clewell, H. (2005) Use of mode of action in risk assessment: past, present, and future. *Regulatory Toxicology and Pharmacology.* **42**, 3–14.

Feynman, R. P. (1999) There's plenty of room at the bottom. In: *The Pleasure of Finding Things Out: The Best Short Works of Richard P. Feynman.* Pages 117–140. Reprint of a talk delivered at Caltech in 1959.

Marchant, G. E. (2002) Toxicogenomics and toxic torts. *TRENDS in Biotechnology.* **20** (8), 329–332.

Meyer, O. (2003) Testing and assessment strategies, including alternative and new approaches. *Toxicology Letters.* **140–141**, 21–30.

Obersdörster, G, *et al.* (2005) Nanotoxicology: an emerging discipline evolving from studies of ultrafine particles. *Environmental Health Perspectives.* **113**, 823–839.

Olson, H. *et al.* (2000) Concordance of the toxicity of pharmaceuticals in humans and animals. *Regulatory Toxicology and Pharmacology.* **32**, 56–67.

Renwick, A. G. (1993) Data-derived safety factors for the evaluation of food additives and environmental contaminants. *Food Additives and Contaminants.* **10**, 275–305.

Rodricks, J. V. (2003) Approaches to risk assessment for macronutrients and amino acids. *Journal of Nutrition.* **133** (6), 2025(s)–2030(s).

Trubo, R. (2005) Endocrine-disrupting chemicals probed as potential pathways to illness. *Journal of the American Medical Association.* **294** (3), 291–294.

World Health Organization. (2002) Risk assessments for salmonella in eggs and broiler chickens. Geneva, World Health Organization.

*Report on Carcinogens: Ninth Annual Report.* (2000) Department of Health and Human, Services, National Institute of Environmental Health Sciences. Research Triangle Park, North, Carolina.

### Chapter 10

Rodricks, J. V. and Reith, S. H. (1998) Toxicological risk assessment in the courtroom. *Regulatory Toxicology and Pharmacology.* **27**, 21–31.

### Chapters 11 and 12

Margolis, H. (1996) *Dealing with Risk: Why the Public and Experts Disagree on Environmental Issues.* Chicago, University of Chicago Press.

Merrill, R. (2001) Regulatory toxicology, in *Casarett and Doull's Toxicology.* 6th edn. New York, McGraw Hill.

Rodricks, J. V. (2001) Some attributes of risk influencing decision-making by public health and regulatory officials. *American Journal of Epidemiology.* **154**, S7–S12.

Slovic, P. (1986) Informing and educating the public about risk. *Risk Analysis.* **6**, 403–415.

Thompson, K. M. (2004) *Risk in Perspective: Insight and Humor in the Age of Risk Management.* Newton Center, MA, AURM Publishers.

Wilson, R. and Crouch, A. C. (2001) *Risk–Benefit Analysis.* Cambridge, MA, Harvard Center for Risk Analysis.

### The Internet

Enormous amounts of information on the topics covered in this book can be found on the Internet. Sites for the important agencies and institutions mentioned in the book are listed. If you are interested in exploring any particular topic in this book, it can be investigated by going to the sites of relevant agencies. Thus, for example, the EPA site will contain information on each of the many categories of chemical products and pollutants the agency regulates.

The agency's methods for assessing risks can also be found. Toxicology information on many specific chemicals is available from the EPA and the Agency for Toxic Substances and Disease Registry, and from the National Toxicology Program. The International Agency for Research on Cancer is a source of information on chemical carcinogens. Sites for other sources of medical and public health information are given in a second section. And in a third section are listed additional sites of high importance. Links to trade associations and advocacy groups with important roles in the topics we cover are provided in the last two sections.

## Agencies and institutions

| | |
|---|---|
| Agency for Toxic Substance and Disease Registry | www.atsdr.cdc.gov |
| Centers for Disease Control | www.cdc.gov |
| Consumer Product Safety Commission | www.cpsc.gov |
| Environmental Protection Agency | www.epa.gov |
| Food and Agriculture Organization | www.fao.org |
| Food and Drug Administration | www.fda.gov |
| Food Safety and Inspection Service | www.fsis.usda.gov |
| International Agency for Research on Cancer | www.iarc.fr/ |
| National Cancer Institute | www.nci.gov |
| National Institutes of Health | www.nih.gov |
| National Institute of Occupational Safety and Health | www.niosh.gov |
| National Toxicology Program | www.ntp-server.niehs.nih.gov/ |
| Occupational Safety and Health Administration | www.osha.gov |
| World Health Organization | www.who.int/en/ |

## Medical and public health information

| | |
|---|---|
| American Cancer Society | www.cancer.org |
| Codex Alimentarius (International food standards) | www.codexalimentarius.net |
| Food Standards Agency, UK | www.food.gov.uk |
| Harvey Project (Human physiology) | www.harveyproject.org |
| The Johns Hopkins Bloomberg School of Public Health | www.jhsph.edu |
| Mayo Clinic (Human diseases) | www.mayoclinic.org |
| The National Academies | www.nationalacademies.org |
| National Library of Medicine | www.nlm.nih.gov |
| | http://sis.nlm.nih.gov/ |
| Physicians Desk Reference (pharmaceuticals) | www.pdrhealth.com/drug_info/ |

### Specific sites of high importance

EPA guidelines for exposure and risk assessment and the agency's integrated
   risk information system (IRIS)
      http://cfpub.epa.gov/ncea/
ATSDR toxicological profiles
      http://www.atsdr.cdc.gov/toxprofiles/
NIOSH databases and information resources
      http://www.cdc.gov/niosh/database.html
FDA toxicology research
      http://www.fda.gov/nctr/index.html
FDA clinical trials
      http://www.fda.gov/oc/gcp/default.htm
FDA food additive safety testing guidelines
      http://www.cfsan.fda/~redbook/red-toca.html
Toxicity testing protocols, international
      www.ich.org
      www.oecd.org
      www.epa.gov/opptsfrs/OPPTS_Harmonized/abguid.html
Occupational health
      http://www.who.int/occupational_health/en/
Environmental health impacts
      http://www.who.int/quantifying_ehimpacts/en/
EU guidelines and regulations for chemicals
      http://europa.eu.int/comm/environment/chemicals/reach.htm
      http://www.efsa.eu.int/
      http://agency.osha.eu.int/OSHA

### Some trade associations

| | |
|---|---|
| American Chemistry Council | www.americanchemistry.com |
| Crop Life of America | www.croplifeamerica.org |
| European Chemical Industry Council | www.cefic.org |
| Food Products Association | www.fpa-food.org/ |
| Pharmaceutical Research and Manufacturers of America | www.pharma.org |

### Some consumer and environmental advocacy groups

| | |
|---|---|
| Center for Science in the Public Interest | www.cspinet.org |
| Environmental Defense | www.environmentaldefense.org |
| Environmental Working Group | www.ewg.org |
| Natural Resources Defense Council | www.nrdc.org |
| Public Citizen | www.citizen.org/index.cfm |

# Index

*Index*